SpringerBriefs in Electrical and Computer Engineering

Control, Automation and Robotics

Series editors

Tamer Başar
Antonio Bicchi
Miroslav Krstic

More information about this series at http://www.springer.com/series/10198

Simona Onori · Lorenzo Serrao
Giorgio Rizzoni

Hybrid Electric Vehicles

Energy Management Strategies

 Springer

Simona Onori
Automotive Engineering Department
Clemson University
Greenville, SC
USA

Lorenzo Serrao
Dana Mechatronics Technology Center
Dana Holding Corporation
Rovereto
Italy

Giorgio Rizzoni
Department of Mechanical and Aerospace
 Engineering and Center for Automotive
 Research
The Ohio State University
Columbus, OH
USA

ISSN 2191-8112   ISSN 2191-8120   (electronic)
SpringerBriefs in Electrical and Computer Engineering
ISSN 2192-6786   ISSN 2192-6794   (electronic)
SpringerBriefs in Control, Automation and Robotics
ISBN 978-1-4471-6779-2   ISBN 978-1-4471-6781-5   (eBook)
DOI 10.1007/978-1-4471-6781-5

Library of Congress Control Number: 2015952754

Springer London Heidelberg New York Dordrecht

Printed on acid-free paper

Springer-Verlag London Ltd. is part of Springer Science+Business Media (www.springer.com)

To my parents, Gianni and Pina

—Simona Onori

To my parents, Salvatore and Silvana

—Lorenzo Serrao

To my family

—Giorgio Rizzoni

Preface

The origin of hybrid electric vehicles dates back to 1899, when Dr. Ferdinand Porsche, then a young engineer at Jacob Lohner & Co, built the first hybrid vehicle [1], the Lohner-Porsche gasoline-electric *Mixte*. After Porsche, other inventors proposed hybrid vehicles in the early twentieth century, but then the internal combustion engine technology improved significantly and hybrid vehicles, much like battery-electric vehicles, disappeared from the market for a long time.

Nearly a century later, hybrid powertrain concepts returned strongly, in the form of many research prototypes but also as successful commercial products: Toyota launched the Prius—the first purpose-designed and -built hybrid electric vehicle—in 1998, and Honda launched the Insight in 1999. What made the new generation of hybrid vehicles more successful than their ancestors was the completely new technology now available, especially in terms of electronics and control systems to coordinate and exploit at best the complex subsystems interacting in a hybrid vehicle. Substantial support to research in this field was provided by government initiatives, such as the US *Partnership for a New Generation of Vehicles* (PNGV) [2], which involved DaimlerChrysler, Ford Motor Company, and General Motors Corporation. PNGV provided the opportunity for many research projects to be carried out in collaborations among the automotive companies, their suppliers, national laboratories, and universities. The material assembled in this book is an outgrowth of the experience that the authors gained while working together at the Ohio State University Center for Automotive Research, one of the PNGV academic labs, which has been engaged in programs focused on the development of vehicle prototypes and on the development of energy management strategies and algorithms since 1995.

Energy management strategies are necessary to achieve the full potential of hybrid electric vehicles, which can reduce fuel consumption and emissions in comparison to conventional vehicles, thanks to the presence of a reversible energy storage device and one or more electric machines. The presence of an additional energy storage device gives rise to new degrees of freedom, which in turn translate into the need of finding the most efficient way of splitting the power demand between the engine and the battery. The energy management strategy is the control layer to which this task is demanded.

Despite many articles on hybrid electric vehicles system, control, and optimization, there has not been a book that systematically discusses deeper aspects of the model-based design of energy management strategies. Thus, the aim of this book is to present a systematic model-based approach and propose a formal framework to cast the energy management problem using optimal control theory tools and language.

The text focuses on the development of model-based supervisory controller when the fuel consumption is being minimized. It does not consider other cost functions, such as pollutant emissions or battery aging. Drivability issues such as noise, harshness, and vibrations are neglected as well as heuristic supervisory controllers design.

The aim is to provide an adequate presentation to meet the ever-increasing demand for engineers to look for rigorous methods for hybrid electric vehicles analysis and design.

We hope that this book will be suitable to educate mechanical and electrical engineering graduate students, professional engineers, and practitioners on the topic of hybrid electric vehicle control and optimization.

Acknowledgments

We are extremely grateful to all our colleagues for the fruitful discussions on the topics discussed in this book. We are also grateful to Springer editorial staff for their support and patience.

August 2015

Simona Onori
Lorenzo Serrao
Giorgio Rizzoni

References

1. Hybrid cars. (Online). Available http://www.hybridcars.com/history/history-of-hybrid-vehicles.html
2. F. Matter, Review of the research program of the partnership for a new generation of vehicles: Seventh report, Washington, DC: The National Academies Press, Tech. Rep. (2001)

Contents

Symbols

t	Time
t_0	Initial time of the optimization horizon
t_f	Final time of the optimization horizon
$x(t)$	State of the optimal control problem
$\mathbf{x}(t)$	State vector, $\mathbf{x}(t) \in \mathbb{R}^n$
$u(t)$	Scalar control of the optimal control problem
$\mathbf{u}(t)$	Control vector, $\mathbf{u}(t) \in \mathbb{R}^p$
$*$	(Relative to the optimal solution)
a_{veh}	Vehicle acceleration
A_f	Frontal area of the vehicle
a	Distance of CG from front axle
b	Distance of CG from rear axle
C_d	Coefficient of aerodynamic drag
c_{roll}	Rolling resistance coefficient
c_{rr0}	Rolling resistance model coefficient (constant term)
c_{rr1}	Rolling resistance model coefficient (speed-dependent term)
C	Electrical capacitance
E_{aero}	Energy dissipated in aerodynamic resistance
E_{batt}	Battery energy
E_{kin}	Kinetic energy
E_{pot}	Potential energy
E_{pwt}	Energy delivered at the wheels by the powertrain
E_{roll}	Energy dissipated in rolling resistance
F_{aero}	Aerodynamic resistance
F_{grade}	Grade force (due to slope)
$F_{inertia}$	Inertial force
F_{roll}	Rolling resistance
F_{trac}	Total tractive force at the wheel-road interface
g_{fd}	Gear ratio (final drive)
g_{fb}	Gear ratio (generic follower/base ratio)

g_{tr}	Gear ratio (transmission)
$G(x,t)$	State constraints
h_{CG}	Height of the center of gravity
$H(\cdot)$	Hamiltonian function
i_{tr}	Gear index (transmission)
I	Current
J	Cost function of optimal control problem
K_{tc}	Capacity factor (in torque converter)
k	Time index in discrete-time problems
L	Instantaneous cost of optimal control problem
$\dot{m}_{elec}$	Instantaneous virtual fuel consumption corresponding to the use of electrical power
$\dot{m}_{eqv}$	Instantaneous equivalent fuel consumption
$\dot{m}_f$	Instantaneous fuel consumption (fuel mass flow rate)
m_f	Total fuel consumption (fuel mass)
M_{veh}	Vehicle mass
MR	Multiplication ratio or torque ratio (in torque converter)
N	Number of elements
P_{acc}	Mechanical power for secondary accessories
$P_{gen,e}$	Electrical power at the generator
$P_{gen,m}$	Mechanical power at the generator
P_{eng}	Mechanical power generated by the internal combustion engine
P_{pto}	Mechanical power for PTO (power take-off) accessories
P_{req}	Total power request by the driver
P_{trac}	Total tractive power at the wheel
Q_{nom}	Nominal charge capacity (of a battery)
Q_{lhv}	Fuel lower heating value
R_0	Electrical resistance
RPM	Rotational speed expressed in revolutions per minute
SR	Speed ratio (in torque converter)
T_b	Base shaft torque (in generic gear set)
T_{brake}	Brake torque at the wheel
T_c	Carrier torque (in planetary gear set)
T_{evt}	Torque at the output of EVT transmission
T_{eng}	Internal combustion engine torque
T_{em}	Electric machine torque
T_f	Follower shaft torque (in generic gear set)
T_{gen}	Electric generator torque
T_{mot}	Electric motor torque
T_{pwt}	Powertrain torque at the wheel
T_p	Pump (impeller) torque (in torque converter)
T_r	Ring torque (in planetary gear set)
T_s	Sun torque (in planetary gear set)
T_t	Turbine torque (in torque converter)

T_{trac}	Total tractive torque at the wheel
U	Admissible control set
V_L	Load voltage at the battery terminals
V_{oc}	Open circuit voltage
v_{veh}	Vehicle speed
Y	Cost to go

BMS	Battery Management System
BSFC	Brake specific fuel consumption
CG	Center of gravity
DP	Dynamic Programming
ICE	Internal combustion engine
PMP	Pontryagin's Minimum Principle
RESS	Rechargeable energy storage system
SOC	State of charge

α	Accelerator pedal position (normalized)
β	Brake pedal position (normalized)
δ	Road grade
η	Efficiency
λ	Co-state variable
μ	Optimal control matrix in dynamic programming
ν	Friction coefficient
ω_b	Base shaft speed (in generic gear set)
ω_c	Carrier speed (in planetary gear set)
ω_{evt}	Speed at the output of EVT transmission
ω_{eng}	Internal combustion engine speed
ω_{em}	Electric machine speed
ω_b	Follower shaft speed (in generic gear set)
ω_{gen}	Electric generator speed
ω_{mot}	Electric motor speed
ω_p	Pump (impeller) speed (in torque converter)
ω_r	Ring speed (in planetary gear set)
ω_s	Sun speed (in planetary gear set)
ω_t	Turbine speed (in torque converter)
ω_{wh}	Wheel speed
Ω_x	Set of admissible states
$\phi(x_f, t_f)$	Terminal cost of optimal control problem
π	Control policy (in dynamic programming)
ρ	Planetary gear ratio
ρ_{air}	Air density
θ	Temperature
ζ	State of energy

Chapter 1
Introduction

1.1 Hybrid Electric Vehicles

Hybrid vehicles are so defined because their propulsion systems are equipped with two energy sources, complementing each other: a high-capacity storage (typically a chemical fuel in liquid or gaseous form), and a lower capacity rechargeable energy storage system (RESS) that can serve as an energy storage buffer, but also as a means for recovering vehicle kinetic energy or to provide power assist. The RESS can be electrochemical (batteries or supercapacitors), hydraulic/pneumatic (accumulators) or mechanical (flywheel) [1]. This dual energy storage capability, in which the RESS permits bi-directional power flows, requires that at least two energy converters be present, at least one of which must also have the ability to allow for bi-directional power flows. *Hybrid electric* vehicles (HEVs), which represent the majority of hybrid vehicles on the road today, use electrochemical batteries as the RESS, and electric machines (one or more) as secondary energy converters, while a reciprocating internal combustion engine (ICE), fueled by a hydrocarbon fuel, serves as the primary energy converter. A fuel cell or other types of combustion engine (gas turbine, external combustion engines) could also serve as the primary energy converter.

The RESS can be used for regenerative braking and also acts as an energy buffer for the primary energy converter, e.g., an ICE, which can instantaneously deliver an amount of power different than what is required by the vehicle load. This flexibility in engine management results in the ability to operate the engine more often in conditions where it is more efficient or less polluting [2, 3]. Other benefits offered by hybridization are the possibility to shut down the engine when it is not needed (such as at a stop or at low speed), and the downsizing of the engine: since the peak power can be reached by summing the output from the engine and from the RESS, the former can be downsized, i.e. replaced with a smaller and less powerful engine, operating at higher average efficiency. *Plug-in* hybrid electric vehicles (PHEVs) allow battery recharge from the electric grid and offer a significant range in pure electric mode. The details of what can actually be accomplished depend on the architecture of the propulsion system and of the vehicle powertrain, as described in the next section.

© The Author(s) 2016
S. Onori et al., *Hybrid Electric Vehicles*, SpringerBriefs in Control,
Automation and Robotics, DOI 10.1007/978-1-4471-6781-5_1

1.2   HEV Architectures

The powertrain of a conventional vehicle is composed by an internal combustion engine, driving the wheels through a transmission that realizes a variable speed ratio between the engine speed and the wheel speed. A dry clutch or hydrodynamic torque converter interposed between engine and transmission decouples the engine from the wheels when needed, i.e., during the transients in which the transmission speed ratio is being modified. All the torque propelling the vehicle is produced by the engine or the mechanical brakes, and there is a univocal relation between the torque at the wheels and the torque developed by the engine (positive) or the brakes (negative).

Hybrid electric vehicles, on the other hand, include one or more electric machines coupled to the engine and/or the wheels [4]. A possible classification of today's vehicles in the market can be given based on internal combustion engine size and electric machine size as shown in Fig. 1.1 [5] and detailed in the following:

1. Conventional ICE vehicles;
2. Micro hybrids (start/stop);
3. Mild hybrids (start/stop + kinetic energy recovery + engine assist);
4. Full hybrids (mild hybrid capabilities + electric launch);
5. Plug-in hybrids (full hybrid capabilities + electric range);
6. Electric Vehicles (battery or fuel cell).

Differences and main characteristics of the different types of vehicles are outlined below [2, 6–8].

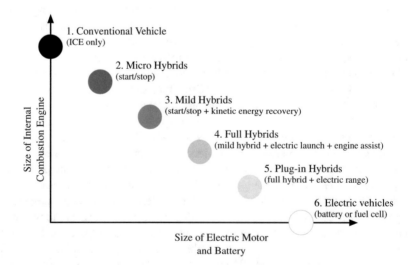

Fig. 1.1 Spectrum of vehicle technologies [5]: pathway of increasing electrification starting with ICE only—powered vehicles, going through different means of vehicle hybridization and ending up with pure electric vehicles powered by batteries or hydrogen fuel cell

1. In conventional vehicles the ICE is the only source of power. For this type of vehicles the total power request at the wheel is entirely satisfied by the ICE.
2. Start–stop systems allow the ICE to shut down and restart to reduce the amount of time spent idling, thus reducing fuel consumption and emissions. This solution is very advantageous for vehicles which spend significant amounts of time waiting at traffic lights or frequently come to a stop in traffic. This feature is present in hybrid electric vehicles, but has also appeared in vehicles which lack a hybrid electric powertrain. Nonelectric vehicles featuring start–stop systems are called micro hybrids.
3. In mild hybrid vehicles generally the ICE is coupled with an electric machine (one motor/generator in a parallel configuration) allowing the engine to be turned off whenever the car is coasting, braking, or stopping. Mild hybrids can also employ regenerative breaking and some level of ICE power assist, but do not have an exclusive electric-only mode of propulsion.
4. Full hybrid electric vehicles run on just the engine, just the battery, or a combination of both. A high-capacity battery pack is needed for battery-only operation during the electric launch. Differently from micro and mild hybrids, where simple heuristic rules are typically used to coordinate the ICE start–stop and power assist functionality, in full hybrid vehicles energy management strategies are needed to fully exploit the benefits of vehicle hybridization, by providing coordination among the actuators in order to minimize fuel consumption.
5. Plug-in hybrid electric vehicles are hybrid vehicles utilizing rechargeable batteries that can be restored to full charge by connecting them to an external electric power source. PHEVs share the characteristics of both full hybrid electric vehicles, having electric motor and an ICE, and of all-electric vehicles, having a plug to connect to the electrical grid.
6. Electric vehicles are propelled only by their on-board electric motor(s), which are powered by a battery (recharged from the power grid) or a hydrogen fuel cell.

In this book, we focus on full hybrid electric vehicles. The number and position of the machines present in full hybrid vehicles define the powertrain architecture, and therefore the performance and capabilities of the hybrid vehicles themselves. HEV architectures can be classified as follows [8]:

- *series*: the engine drives a generator, producing electrical power which can be summed to the electrical power coming from the RESS and then transmitted, via an electric bus, to the electric motor(s) driving the wheels;
- *parallel*: the power summation is mechanical rather than electrical: the engine and the electric machines (one or more) are connected with a gear set, a chain, or a belt, so that their torque is summed and then transmitted to the wheels;
- *power split*: the engine and two electric machines are connected to a power split device (usually a planetary gear set), thus the power from the engine and the electric machines can be merged through both a mechanical and an electrical path, thus combining series and parallel operation;

- *series/parallel*: the engagement/disengagement of one or two clutches allows to change the powertrain configuration from series to parallel and vice versa, thus allowing the use of the configuration best suited to the current operating conditions.

The series architecture has the advantage of requiring only electrical connections between the main power conversion devices. This simplifies some aspects of vehicle packaging and design. Also, having the engine completely disconnected from the wheels gives great freedom in choosing its load and speed, thus making it possible for the engine to operate at the highest possible efficiency. On the other hand, series hybrids require two energy conversions (mechanical to electrical in the generator, and electrical to mechanical in the motor), which introduce losses, even in cases when a direct mechanical connection of the engine to the wheels would actually be more efficient, overall. For this reason, there are conditions in which a series hybrid vehicle consumes more fuel than its conventional counterpart: for example, in highway driving. Further, one of the two electromechanical energy converters must be sized to support the maximum power requirements of the vehicle, since it is the primary source of motive power. The parallel architecture does not have this problem; however, unless significantly oversized, the electric motors are less powerful than those used in a series hybrid (because not all the mechanical power flows through them), thus reducing the potential for regenerative braking; also, the engine operating conditions cannot be determined as freely as in a series hybrid architecture, because the engine speed is mechanically related (via the transmission) to the vehicle velocity. Power split and series/parallel architectures (which can be realized in different ways) are the most flexible, and give a higher degree of control of the operating conditions of the engine than the parallel architecture while applying the double energy conversion typical of series operation only to a fraction of the total power flow, thus reducing overall losses [3, 8].

1.3   Energy Analysis of Hybrid Electric Vehicles

Energy analysis is essential to understand why hybrid electric vehicles are beneficial from the efficiency point of view, and to appropriately design and assess energy management strategies. Consider the case of a conventional vehicle: the combustion engine, which converts the chemical energy in the fuel to mechanical energy, generates all the power needed during a trip. The mechanical power generated by the engine is used for moving all driveline components, driving accessories (power steering, alternator, air conditioning ...), and, of course, moving the vehicle. Given the driver's input (accelerator and brake pedals) and the driving conditions (speed, road surface, etc.), the operating condition of the engine (speed and torque) is determined by a single degree of freedom, i.e., the choice of the transmission gear ratio. The *power management* strategy is the choice of this ratio. In hybrid electric vehicles, instead, the total power demand is satisfied by summing together the outputs of the engine (thermal path or fuel path) and of the battery or other storage devices (electric path). The ratio of the power flows generated by each path constitutes an additional degree of freedom that permits optimization of the engine operating conditions to achieve improvements in efficiency and fuel economy. In addition, the electric motors are reversible and can produce negative torque. Thus, they can replace or supplement the mechanical brakes as a means to decelerate the vehicle, with the benefit of acting like generators and producing electrical energy, which can be stored in batteries on board of the vehicle for later use. This operation, known as *regenerative braking*, may substantially improve the overall efficiency over an extended time period. The additional freedom afforded by a hybrid architecture makes the use of a power management strategy necessary, both over a short time horizon, to recover braking energy and to guarantee performance and instantaneous fuel economy, as well as over a long time horizon, to guarantee that the RESS has sufficient energy in store when needed, and that fuel economy benefits are achieved. Hence, the need for an *energy management* strategy arises, which extends the power management (instantaneous) with considerations based on a longer time horizon, keeping into account the amount of energy stored in the vehicle.

The energy management strategy determines at each instant the power repartition between the engine and the RESS, according to instantaneous constraints (e.g., generating the total power output requested by the driver), global constraints (e.g., maintaining the RESS energy level within safety limits) and global objectives (e.g., minimizing the fuel consumption during a trip).

1.4   Book Structure

The objective of this book is to illustrate optimization-based methods to design a high-level energy management strategy for hybrid electric vehicles, based on optimal control theory tools. Chapter 2 provides an overview on control-oriented modeling approaches and methods that can be used for energy management design and testing; Chap. 3 defines the role of energy management system in the overall vehicle control architecture and introduces the energy management problem in a rigorous way. Chapter 4 presents the Dynamic Programming (DP) algorithm and Chap. 5 Pontryagin's minimum principle (PMP), two optimal control methods, applied to the HEV energy management problem to obtain the theoretical optimal solution, and as such only applicable offline; in Chap. 5 the relation between the DP and PMP solutions is presented. Chapter 6 describes the Equivalent Fuel Consumption Minimization Strategy (ECMS), and discusses its equivalence to the PMP solution. A family of online causal suboptimal control strategies derived from PMP/ECMS is introduced in Chap. 7. Here, adaptive methods to update the control parameter used in the PMP/ECMS are discussed, which result into suboptimal real-time implementable strategies. Finally, Chap. 8 presents two case studies to demonstrate with practical examples the application of the modeling techniques and the implementation in simulation of optimal control strategies.

References

1. W. Liu, *Introduction to Hybrid Vehicle System Modeling and Control* (Wiley, Hoboken, 2013)
2. L. Guzzella, A. Sciarretta, *Vehicle Propulsion Systems: Introduction to Modeling and Optimization* (Springer, Berlin, 2013)
3. G. Rizzoni, H. Peng, Hybrid and electric vehicles: the role of dynamics and control. ASME Dyn. Syst. Control Mag. **1**(1), 10–17 (2013)
4. C.C. Chan, The state of the art of electric and hybrid vehicles. Proc. IEEE **90**(2), 245–275 (2002)
5. A.A. Pesaran, Choices and requirements of batteries for EVs, HEVs, PHEVs, in NREL/PR-5400-51474 (2011)
6. F. An, F. Stodolsky, D. Santini, Hybrid options for light-duty vehicles, in SAE Technical Paper No. 1999-01-2929 (1999)
7. F. An, A. Vyas, J. Anderson, D. Santini, Evaluating commercial and prototype HEVs, in SAE Technical Paper No. 2001-01-0951 (2001)
8. J.M. Miller, *Propulsion Systems for Hybrid Vehicles* (The Institution of Electrical Engineers, London, 2003)

Chapter 2
HEV Modeling

2.1 Introduction

The objective of the energy management control is to minimize the vehicle fuel consumption, while maintaining the battery state of charge around a desired value. To this end, modeling for energy management may have two scopes: creating plant simulators to which an energy management strategy is applied for testing and development, or creating embedded models that are used to set up analytically and/or solve numerically the energy management problem. Plant models tend to be more accurate and computationally heavy than embedded control models. The main objective in both cases is to reproduce the energy flows within the powertrain and the vehicle, in order to obtain an accurate estimation of fuel consumption and battery state of charge, based on the control inputs and the road load. In some applications, other quantities may be of interest, such as thermal flows (temperature variation in engine, batteries, after-treatment, etc.), battery aging, pollutant emissions, etc.

This chapter provides a concise overview of the modeling issues linked to the development and simulation of energy management strategies. The reader is referred to more specialized works for further details (e.g., [1]). Efficiency considerations are at the basis of the models described, which are suited for preliminary analysis and high-level energy management development.

2.2 Modeling for Energy Analysis

Because of the losses in the powertrain, the net amount of energy produced at the wheels is smaller than the amount of energy introduced into the vehicle from external sources (e.g., fuel). Conversion losses take place when power is transformed into a different form (e.g., chemical into mechanical, mechanical into electrical, etc.). Similarly, when power flows through a connection device, friction losses and other inefficiencies reduce the amount of power at the device output. Energy losses in powertrain components are usually modeled using efficiency maps, i.e., tables that contain efficiency data as a function of the operating conditions (for example, the output torque and the rotational speed of the engine). Maps are built experimentally

© The Author(s) 2016   7
S. Onori et al., *Hybrid Electric Vehicles*, SpringerBriefs in Control,
Automation and Robotics, DOI 10.1007/978-1-4471-6781-5_2

as a set of stationary points, i.e., letting the component reach a steady-state operating condition and measuring power input and output (and/or power dissipation) in that condition. Because of this procedure, efficiency maps may not be accurate during transients. Despite this, the approach is widely used because it allows to generate simple models capable of being evaluated quickly when implemented in computer code, and validation results [2] show that the accuracy of such models can be very good for estimating fuel consumption and energy balance, as most of the energy content is associated with the slower system dynamics [3].

The vehicle fuel consumption for a prescribed driving cycle can be estimated using a *backward* or a *forward* modeling approach. The backward, quasi-static approach is based on the assumption that the prescribed driving cycle is followed exactly by the vehicle. The driving cycle is subdivided in small time intervals, during which an average operating point approach is applied, assuming that speed, torque, and acceleration remain constant: this is equivalent to neglecting internal powertrain dynamics and taking average values of all variables during the selected sampling time, which is therefore longer than typical powertrain transients (e.g., engine dynamics, gear shifting), and of the same order of magnitude of vehicle longitudinal dynamics and driving cycle variations. Each powertrain component is modeled using an efficiency map, a power loss map, or a fuel consumption map: these give a relation between the losses in the component and the present operating conditions (averaged during the desired time interval).

The forward, dynamic approach is based on a first-principles description of each powertrain component, with dynamic equations describing the evolution of its state. The degree of modeling detail depends on the timescale and the nature of the phenomena that the model should predict. In the simplest case, the same level of detail as the quasi-static approach can be applied, but the evolution of vehicle speed is computed as the result of the dynamic simulation and not prescribed a priori.

2.3 Vehicle-Level Energy Analysis

By vehicle-level energy analysis, we refer to the case in which the vehicle is considered as a point mass and its interaction with the external environment is studied, in order to compute the amount of power and energy needed to move it with specified speed. This high-level approach is useful to develop an understanding of the vehicle longitudinal dynamics and of the energy characteristics of hybrid vehicles.

2.3.1 Equations of Motion

If a vehicle is considered as a mass point, its motion equation can be written from the equilibrium of forces shown in Fig. 2.1:

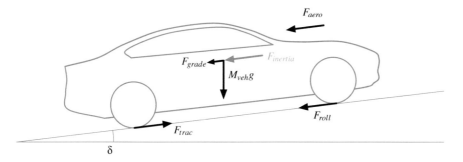

Fig. 2.1 Forces acting on a vehicle

$$M_{veh}\frac{dv_{veh}}{dt} = F_{inertia} = F_{trac} - F_{roll} - F_{aero} - F_{grade}, \qquad (2.1)$$

where M_{veh} is the effective vehicle mass, v_{veh} is the longitudinal vehicle velocity, $F_{inertia}$ is the inertial force, $F_{trac} = F_{pwt} - F_{brake}$ is the tractive force generated by the powertrain and the brakes at the wheels,[1] F_{roll} is the rolling resistance (friction due to tire deformation and losses), F_{aero} the aerodynamic resistance, F_{grade} the force due to road slope.

The aerodynamic resistance is expressed as

$$F_{aero} = \frac{1}{2}\rho_{air}A_f C_d v_{veh}^2, \qquad (2.2)$$

where ρ_{air} is the air density ($1.25\,\text{kg/m}^3$ in normal conditions), A_f the vehicle frontal area, C_d the aerodynamic drag coefficient.

The rolling resistance force is usually modeled as [1]

$$F_{roll} = c_{roll}(v_{veh}, p_{tire}, \ldots)M_{veh}g\cos\delta, \qquad (2.3)$$

where g is the gravity acceleration, δ the road slope angle (so that $M_{veh}g\cos\delta$ is the vertical component of the vehicle weight), and c_{roll} is a rolling resistance coefficient which is, in principle, a function of vehicle speed, tire pressure p_{tire}, external temperature, etc. In most cases, c_{roll} is assumed to be constant, or to be an affine function of the vehicle speed:

$$c_{roll} = c_{r0} + c_{r1}v_{veh}. \qquad (2.4)$$

---

[1]This is the sum of the forces acting on the individual wheels. For each wheel, it represents the net torque acting on the wheel divided by the effective tire radius. Note that the tire radius is assumed here to be equal to the nominal tire radius, but it can be very different from this value during dynamic transient maneuvers, which are not considered in this book. See a vehicle dynamics textbook for more details on the modeling of ground/tire forces (see, e.g., [4]).

Table 2.1 Typical values of vehicle-dependent parameters for longitudinal vehicle dynamics models

Parameter	Compact car	Full-size car	SUV
M_{veh}	1200–1500 kg	1700–2000 kg	1900–2200 kg
C_d	0.3–0.35	0.28–0.33	0.32–0.38
A_f	1.3–1.7 m²	1.8–2.2 m²	2–2.5 m²
c_{roll}	0.01–0.03	0.01–0.03	0.01–0.03

The order of magnitude of c_{roll} is 0.01–0.03 (for a light vehicle on normal road surface), which means that the rolling resistance is 1–3 % of the vehicle weight (depending on vehicle, soil, tires and tire pressure, temperature, etc.).

The grade force is the horizontal component of the vehicle weight, which opposes (or facilitates) vehicle motion only if the vehicle is moving uphill (or downhill):

$$F_{grade} = M_{veh}g \sin \delta. \tag{2.5}$$

These basic equations represent the starting point for vehicle modeling, and can be sufficiently accurate if the parameters are correctly identified. Typical values of the vehicle-level parameters are listed in Table 2.1.

2.3.2 Forward and Backward Modeling Approaches

Equation (2.1) can be rearranged to calculate the tractive force that the powertrain needs to produce, given the acceleration (inertial force $F_{inertia}$):

$$F_{trac} = F_{pwt} - F_{brake} = F_{inertia} + F_{grade} + F_{roll} + F_{aero}. \tag{2.6}$$

The different form of (2.1) and (2.6) corresponds to the forward and backward modeling approaches: in (2.1), the vehicle acceleration $\frac{dv_{veh}}{dt}$ is computed as a consequence of the tractive force generated by the powertrain (and obviously the external resistance terms), and the speed is then obtained by integration of the acceleration; this is the *forward* approach, which reproduces the physical causality of the system. On the other hand, in the *backward* approach modeled by (2.6), force follows velocity and the tractive force is calculated starting from the inertia force: in this case, it is assumed that the vehicle is following a prescribed velocity (and acceleration) profile, and F_{trac} represents the corresponding force that the powertrain must supply.

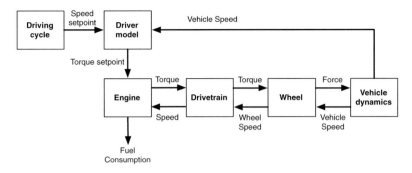

Fig. 2.2   Information flow in a forward simulator

The forward approach is the option typically chosen in most simulators; it is characterized by the information flow as shown in Fig. 2.2. For example, in the case of a hybrid vehicle forward simulator, the desired speed (from the cycle inputs) is compared to the actual vehicle speed, and braking or throttle commands are generated using a driver model (e.g., a PID speed controller) in order to follow the imposed vehicle profile. This driver command is an input to the supervisor block that is responsible of issuing the actuators setpoints (engine, electric machines, and braking torques) to the rest of the powertrain components, which ultimately produce a tractive force. Finally, the force is applied to the vehicle dynamics model, where the acceleration is determined with (2.1), taking into account the road load information [5].

In a backward simulator, instead (see Fig. 2.3), no driver model is necessary, since the desired speed is a direct input to the simulator, while the engine torque and fuel consumption are outputs. The simulator determines the net tractive force to be applied based on the velocity, payload, and grade profiles, along with the vehicle characteristics. Based on this information, the torque that the powertrain should apply is calculated, and then the torque/speed characteristics of the various powertrain components are taken into account in order to determine the engine operating conditions and, finally, the fuel consumption.

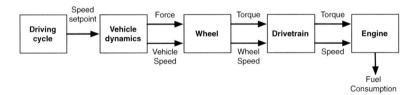

Fig. 2.3   Information flow in a backward simulator

Both the forward and backward simulation approaches have their relative strengths and weaknesses. Fuel economy simulations are typically conducted over predetermined driving cycles, and therefore using a backward simulator ensures that each different simulation exactly follows this profile, which guarantees consistency of simulation results. By contrast, a forward simulator may not exactly follow the trace, as it introduces a small error between the actual and the desired signal. Proper tuning of the driver block can reduce the differences, whereas the backward version keeps the error at zero without any effort. On the other hand, a backward simulation assumes that the vehicle and powertrain are capable of following the speed trace, and does not account for limitations of the powertrain actuators in computing the vehicle speed, which is predetermined. This poses the problem of evaluating demanding cycles which may require more power than the powertrain can provide. A forward simulation does not have this issue, because the speed is computed from the torque/force output, which can be saturated according to the powertrain limitations. For this reason, forward simulation can also be used for acceleration tests and in general for testing the behavior of the system at saturation. In addition, forward simulators are implemented according to physical causality and, if their level of detail is appropriate, can be used for development of online control strategies, while a backward simulator is suited for preliminary screening of energy management strategies. It is possible to combine the advantages of both modeling approaches, i.e., the accurate reproduction of a cycle by a backward simulation and the ability to capture powertrain limitations of a forward simulator. A solution, represented in Fig. 2.4, consists in using a forward simulator in which the driver model (speed controller) uses a backward vehicle model to compute the torque setpoints to be applied: in this way, the resulting speed profile will match exactly the reference cycle, if this does not

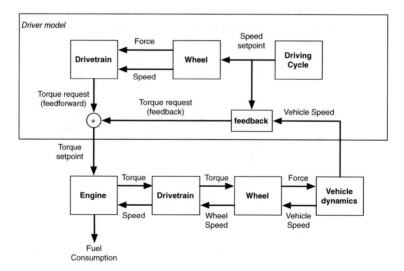

Fig. 2.4 Information flow in a backward–forward simulator

saturate the powertrain capacity, but will be appropriately saturated when needed since it goes through a forward powertrain model. A feedback term should also be added, in order to recover speed deviation due to powertrain saturation (or to possible mismatches between the backward and forward models).

2.3.3 Vehicle Energy Balance

Fuel consumption evaluation is conducted by analyzing the energy flows in the powertrain and identifying the areas in which saving can be introduced. From (2.6) the inertial force $F_{inertia}$ is positive when the vehicle is accelerating, and negative during deceleration; the grade force F_{grade} is positive when the vehicle is driven uphill and negative when it is going downhill; the rolling (F_{roll}) and aerodynamic (F_{aero}) resistances are always positive (for a vehicle moving in forward direction).

The forces F_{roll} and F_{aero} are dissipative, since they always oppose the motion of the vehicle, while the inertial and grade forces are conservative, being only dependent on the vehicle state (respectively velocity and altitude). Thus, part of the tractive force generated by the powertrain increases the kinetic and potential energy of the vehicle (by accelerating it and moving it uphill), and part is dissipated in rolling and aerodynamic resistances. When the vehicle decelerates or drives downhill, its potential and kinetic energy must be dissipated: rolling and aerodynamic resistances contribute to dissipating part of the vehicle energy, but for faster deceleration the mechanical brakes must be used. Thus, ultimately, all the energy that the powertrain produces is dissipated in these three forms: rolling resistance, aerodynamic resistance, and mechanical brakes. The net variation of kinetic energy is always zero between two stops (since initial speed and final speed are both zero), and the variation of potential energy only depends on the difference of altitude between the initial and ending point of the trip considered.

Multiplying all terms of (2.6) by the vehicle speed (v_{veh}) the following balance of power is obtained:

$$P_{trac} = P_{inertia} + P_{grade} + P_{roll} + P_{aero}. \tag{2.7}$$

The term P_{trac} represents the tractive power at the wheels, either positive or negative. Positive P_{trac} is generated by the powertrain to propel the vehicle, while negative P_{trac} (corresponding to deceleration) can be obtained using the powertrain, the brakes or both. In conventional vehicles, the amount of negative power that the powertrain can absorb is rather limited: it consists in friction losses in the various components and pumping losses in the engine. In hybrid electric vehicles, the amount of negative power is much higher, since the electric traction machines are reversible and can be used for deceleration as well as acceleration.

The term $P_{inertia} = M_{veh}\dot{v}_{veh}v_{veh}$ represents the amount of power needed just to accelerate the vehicle (without considering the losses); the terms $P_{roll} = F_{roll}v_{veh}$ and $P_{aero} = F_{aero}v_{veh}$ are the amount of power needed to overcome the rolling and aerodynamic resistances respectively; and $P_{grade} = F_{grade}v_{veh}$ is the power that goes into overcoming a slope (or, if the slope is negative and the vehicle is going downhill, it is the power that accelerates the vehicle and, when excessive, must be dissipated to prevent undesired acceleration).

If the terms that appear in (2.7) are integrated over the duration of a trip (time interval $[t_0 \ t_f]$), the following energy balance is obtained:

$$E_{trac} = \int_{t_0}^{t_f} P_{trac}dt = E_{kin} + E_{pot} + E_{roll} + E_{aero}, \qquad (2.8)$$

where the individual terms are:

$$E_{kin} = \int_{t_0}^{t_f} P_{inertia}dt = M_{veh}\int_{t_0}^{t_f} v_{veh}(t)\dot{v}_{veh}(t)dt; \qquad (2.9a)$$

$$E_{pot} = \int_{t_0}^{t_f} P_{grade}dt = M_{veh}g\int_{t_0}^{t_f} v_{veh}(t)\sin\delta(t)dt; \qquad (2.9b)$$

$$E_{roll} = \int_{t_0}^{t_f} P_{roll}dt = M_{veh}g\int_{t_0}^{t_f} c_{roll}v_{veh}(t)\cos\delta(t)dt; \qquad (2.9c)$$

$$E_{aero} = \int_{t_0}^{t_f} P_{aero}dt = \frac{1}{2}\rho_{air}A_fC_d\int_{t_0}^{t_f} v_{veh}(t)^3dt. \qquad (2.9d)$$

Note that the integral of the inertial power $P_{inertia}$ is the variation of kinetic energy E_{kin}, and the integral of the grade power P_{grade} is the variation of potential energy E_{pot}. Each energy term is the product of two terms: one representing vehicle parameters (mass, resistance coefficients), which are independent of the driving cycle, and the other representing driving cycle information, independent of the vehicle characteristics and only function of the velocity profile[2] $v_{veh}(t)$.

The relative amount of rolling resistance, aerodynamic resistance, and brake energy defines the characteristics of a driving cycle. In particular, the potential for energy recovery using regenerative braking is equal to the amount of kinetic and potential energy that needs to be dissipated, minus the quantity that is dissipated because of rolling and aerodynamic resistance. Thus, a urban driving cycle with frequent accelerations and decelerations at low speed (where the resistances are lower) presents more potential for energy recovery than a highway cycle in which the speed is approximately constant and the losses due to aerodynamic resistance represent the major component of the power requested by the vehicle.

To better understand this concept, it is useful to look separately at the energy balance during acceleration ($\dot{v}_{veh} \geq 0$) and deceleration ($\dot{v}_{veh} < 0$), i.e., compute the integrals above by summing over different sections of the driving cycle. Let us denote with the superscript $^+$ the energy values computed by considering only the instants in which $\dot{v}_{veh} \geq 0$, and with the superscript $^-$ those relative to the instants in

---

[2]An exception is the rolling resistance contribution E_{roll}, because the coefficient c_{roll} may, in general, depend on vehicle speed as well as vehicle and tire characteristics.

which $\dot{v}_{veh} < 0$ (i.e., the integrals (2.9a, 2.9b, 2.9c, 2.9d) are split into two domains, according to the sign of $\dot{v}_{veh}$).

The kinetic energy in the two cases is equal, but with opposite sign:

$$E_{kin}^- = -E_{kin}^+ \tag{2.10}$$

because the net variation of kinetic energy is zero during the entire cycle, and its variation is positive each time $\dot{v}_{veh} > 0$, and negative each time that $\dot{v}_{veh} < 0$.

The amount of energy that the powertrain must deliver during acceleration is thus:

$$E_{pwt}^+ = E_{roll}^+ + E_{aero}^+ + E_{pot}^+ + E_{kin}^+, \tag{2.11}$$

that is, the energy provided by the powertrain is spent to: accelerate the vehicle (increase its kinetic energy by E_{kin}^+); move it at a higher level (E_{pot}^+); and overcome dissipative resistances (E_{roll}^+ and E_{aero}^+). However, in the course of a complete trip (vehicle starting from standstill and coming to a stop at the end), the net variation of kinetic energy is zero. Therefore, the same amount of kinetic energy produced during acceleration (E_{kin}^+) must be removed from the vehicle during deceleration.

When the vehicle decelerates, it needs to dissipate the entire amount of kinetic energy accumulated during acceleration. The dissipative resistances contribute to this, since they tend to slow down the vehicle. However, the amount of kinetic energy to dissipate during deceleration may be higher than the sum of rolling and aerodynamic resistance; in this case, the vehicle must be decelerated using additional actuators, for example using mechanical brakes or, in a hybrid vehicle, producing negative torque with electric traction motors, thus recuperating (some of) the energy. The amount of energy available for regeneration, $E_{regen,pot}$, is the total vehicle energy cumulated during acceleration (kinetic and potential) minus the losses during the deceleration phase, given by dissipative losses (rolling resistance and aerodynamic drag) and by the increase of potential energy (E_{pot}^-)[3]:

$$E_{regen,pot} = E_{kin}^+ + E_{pot}^+ - E_{roll}^- - E_{aero}^- - E_{pot}^- \tag{2.12}$$

The diagram in Fig. 2.5 shows graphically this concept: proceeding from left to right, losses are subtracted to compute the energy available at each stage.

2.3.4 Driving Cycles

As implied in the previous section, the advantages of hybrid vehicles depend on how the vehicle is used. In particular, the hybridization advantages consist essentially in

---

[3]In other words, if the vehicle is decelerating uphill, part of its kinetic energy is lost to overcome the gravity; downhill, on the other hand, the gravitational force will increase the amount of energy to be regenerated.

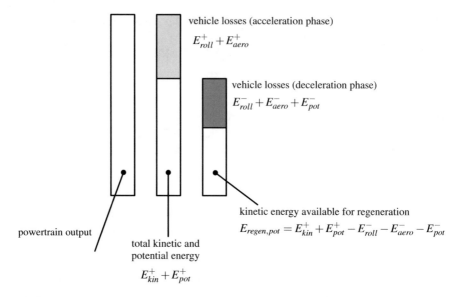

vehicle losses (acceleration phase)

$$E^+_{roll} + E^+_{aero}$$

vehicle losses (deceleration phase)

$$E^-_{roll} + E^-_{aero} + E^-_{pot}$$

kinetic energy available for regeneration

$$E_{regen,pot} = E^+_{kin} + E^+_{pot} - E^-_{roll} - E^-_{aero} - E^-_{pot}$$

powertrain output

total kinetic and
potential energy

$$E^+_{kin} + E^+_{pot}$$

Fig. 2.5 Vehicle energy balance (bar length represents energy)

recovering potential and kinetic energy that would otherwise be dissipated in the
brakes, and in operating the engine in its highest-efficiency region. If the engine had
a constant efficiency and the vehicle drove at constant speed on a flat road, there
would be no advantage in a hybrid electric configuration.

A *driving cycle* represents both the way the vehicle is driven during a trip and the
road characteristics. In the simplest case, it is defined as a time history of vehicle speed
(and therefore acceleration) and road grade. Together with the vehicle characteristics,
this completely defines the road load, i.e., the force that the vehicle needs to exchange
with the road during the driving cycle.

As pointed out in Sect. 2.3.3, each term in the energy balance is a function of
both the driving cycle (speed, acceleration, grade) and the vehicle (mass, frontal
area, coefficients of aerodynamic and rolling resistance). For this reason, the fuel
consumption of a vehicle must always be specified in reference to a specific driving
cycle. On the other hand, given a driving cycle, the absolute value of the road load and
also the *relative* magnitude of its components depend on the vehicle characteristics.

The necessity for a standard method to evaluate emissions and fuel consumption
of all vehicles on the market, and to provide a reliable basis for their comparison,
led to the introduction of a reduced number of regulatory driving cycles: any vehicle
sold must be tested, according to detailed procedures, using one or more of these
standard cycles, which are different for each world region.

Examples of standard cycles are shown in Fig. 2.6, which also include a basic
energy analysis comparison.

These driving cycles are designed to be representative of urban and extra-urban
driving conditions. The Japan 10–15 and European cycle (NEDC) are synthetic,

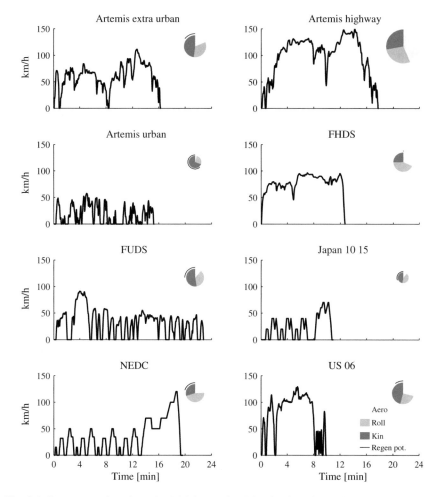

Fig. 2.6 Some examples of standard driving cycles. The pie chart shows the relative amount of the energy terms E_{kin}^+, E_{aero}^+, E_{roll}^+, as well as the amount of kinetic energy that can be recovered according to (2.12). The pie surface is proportional to the total cycle energy E_{pwt}^+ defined by (2.11). Energies are computed with the vehicle data of Table 8.1

while the others reproduce measures of vehicle speed in actual roads. However, with the exception of US 06, the acceleration levels are well below the capabilities of any modern car, therefore the fuel consumption results are typically optimistic and unable to reproduce real-world driving conditions.

The regulatory cycles should be considered a standard comparison tool and not as representative of actual operating conditions. In fact, it is not possible to predict how a vehicle will be driven, since each vehicle has a different usage pattern and each driver his or her own driving style. In order to obtain more realistic estimations of real-world fuel consumption for a specific vehicle, vehicle manufacturers may develop their own testing cycles.

2.4 Powertrain Components

This section contains a description of models of the principal powertrain components suitable for energy flow modeling, neglecting component dynamics. Detailed behavioral models accurately accounting for dynamic effect are beyond the objectives of this book and can be found in specialized works.

2.4.1 Internal Combustion Engine

The following modeling approaches can be used for an internal combustion engine, in order of increasing complexity:

1. Static map;
2. Static map and lumped-parameter dynamic model;
3. Mean-value model;
4. One-dimensional fluid-dynamic model;
5. Three-dimensional fluid-dynamic model (finite-element).

The latter two approaches are necessary only for detailed studies focused on the engine subsystem, while the first three methods can be useful in models in which the engine is seen as part of a more comprehensive system (powertrain or vehicle) and as such can be employed in energy management simulators (map models) or powertrain control strategies (map with lumped-parameter dynamics or mean-value models).

The static map approach assumes the engine to be a perfect actuator, which responds immediately to the commands; the fuel consumption is computed using a map (table) as a function of the engine speed and torque, both of which are assumed to be known. In particular, the torque is typically a control input for the engine, while the speed is a measured input and derives from the coupling to the rest of the powertrain. A curve that gives the maximum engine torque as a function of the current speed is also present in this kind of models to ensure that the torque does not exceed the limits of the engine. Figure 2.7 shows the typical engine map information with fuel consumption or iso-efficiency contours, the maximum torque curve, and the optimal operation line (OOL), i.e., the combination of torque and speed that provide the maximum efficiency for any given power output. The OOL information is often used in designing heuristic energy management strategies, as a target for the engine operating points.

The map-based model can be modified to include dynamic limitations in the torque output, i.e., a delay between the commanded torque and the actual torque generated,

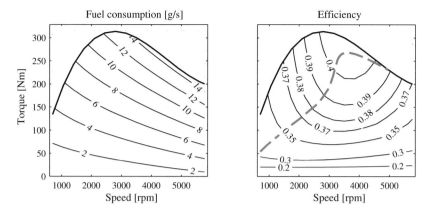

Fig. 2.7 Example of engine fuel consumption map and efficiency map (with optimal operation line, OOL, in *dashed-line*)

by coupling it to a transfer function representing air/fuel dynamics and, possibly, to an inertia representing the crankshaft dynamics.

2.4.2 Torque Converter

The torque converter is a fluid coupling device that is used to transmit motion from the engine to the transmission input shaft. It is capable of multiplying the engine torque (acting as a reduction gear), and, unlike most other mechanical joints, provides extremely high damping capabilities, since all torque is transmitted through fluid-dynamic forces rather than friction or pressure. It is traditionally used in vehicles with automatic transmissions as a launching device, because it allows for large speed differences between its two shafts while multiplying the input torque.

A torque converter (Fig. 2.8) is composed by three co-axial elements: a pump (also called impeller), connected to the engine shaft, a turbine, connected to the transmission, and a stator in between. The fluid in the torque converter is moved by

Fig. 2.8 Schematic representation of a torque converter

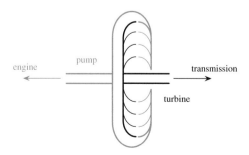

the pump because of engine rotation, drags the turbine, and therefore transmits torque to the transmission. The torque at the turbine is multiplied with respect to the pump torque (i.e., the engine torque), thanks to the presence of the stator which modifies the flow characteristics inside the converter. The torque multiplication increases with the speed difference between the pump and the turbine; at steady state, the two elements rotate at the same speed and the torque multiplication factor is unitary.

The torque converter model is based on tabulated characteristics of torque ratio and capacity factor versus speed ratio. The speed ratio is

$$SR = \frac{\omega_t}{\omega_p}, \tag{2.13}$$

where ω_t is the turbine speed and ω_p the pump speed. The torque ratio or multiplication ratio is

$$MR = \frac{T_t}{T_p}, \tag{2.14}$$

with T_t and T_p the turbine and pump torque respectively. The capacity factor, which is a measure of how much torque the torque converter can transmit, is defined as

$$K_{tc} = \frac{\omega_p}{\sqrt{T_p}}. \tag{2.15}$$

As an alternative to the capacity factor, the torque at 2000 rpm (MP_{2000}) is sometimes used to characterize the torque capacity; it is related to the capacity factor as follows:

$$MP_{2000} = \frac{2000^2}{K_{tc}^2}, \tag{2.16}$$

where K_{tc} must be expressed in units of $\frac{\text{RPM}}{\sqrt{\text{Nm}}}$.

Examples of characteristic curves of a torque converter are shown in Fig. 2.9. The map can be replaced by an analytical model, the Kotwicki model [6], based on curve fitting.

2.4.3 Gear Ratios and Mechanical Gearbox

Gearings are purely mechanical components, with no control, that change the speed and torque transmitted between two shafts without altering the power flow. In practice, however, losses due to friction occur and reduce the output power with respect to the input power.

The simplest model for a gearing only accounts for the speed and torque ratios, without considering the losses due to friction. Indicating with the subscripts b and f the base and follower shaft (see Fig. 2.10), and with $g_{fb} = \frac{N_b}{N_f}$ the transmission ratio

Fig. 2.9 Example of
a torque converter map

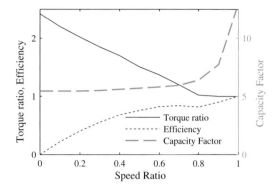

Fig. 2.10 Schematic
representation of a gear
coupling

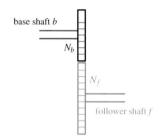

(N is the number of teeth of each gear), the lossless gear model is:

$$\begin{cases} \omega_f = g_{fb}\omega_b, \\ T_f = \dfrac{1}{g_{fb}}T_b. \end{cases} \tag{2.17}$$

For energy analysis and in general for more accurate predictions, a lossy gear model is introduced, which takes into account power losses. Given that the speed ratio is fixed, being given by kinematic constraints, the speed equation remains the same as the lossless model, while the power loss means a reduction of the torque at the output shaft, described using the gear efficiency η_{fb}:

$$T_f = \begin{cases} \dfrac{\eta_{fb}}{g_{fb}}T_b & \text{if } P_b = T_b \cdot \omega_b \geq 0, \\ \dfrac{1}{\eta_{fb}\cdot g_{fb}}T_b & \text{if } P_b = T_b \cdot \omega_b < 0. \end{cases} \tag{2.18}$$

with the convention that power flow is positive when going from b to f, i.e., when b is the input shaft. The power loss is always positive and is calculated as

$$P_{loss} = \begin{cases} \omega_b T_b(1 - \eta_{fb}) & \text{if } P_b = T_b \cdot \omega_b \geq 0, \\ \omega_f T_f(1 - \eta_{fb}) & \text{if } P_b = T_b \cdot \omega_b < 0. \end{cases} \tag{2.19}$$

Functionally, a gearbox is a gearing whose transmission ratio (and possibly other characteristics, such as efficiency) can change dynamically. The simplest model for a gearbox consists in a lossy gear with variable gear ratio; the efficiency can be assumed constant or variable with gear ratio, speed, and input torque. This model captures the essential functionality common to manual gearboxes and automatic transmissions, and can be used for both cases. A complete transmission model with several degrees of freedom (considering all the gears, coupling and actuators) is more suited for drivability studies.

2.4.4   Planetary Gear Sets

Planetary gear sets are composed by three rotating elements (sun, carrier, and ring) which are connected by internal gears (planets); stopping one of the three shafts generates a fixed gear ratio between the remaining two. Planetary gears are commonly used in traditional automatic transmissions because they allow for compact construction and smooth gear transition. They are often present in hybrid electric vehicles to realize electrically variable transmissions (EVTs) by connecting the engine and two electric machines to the three shafts of the gear set.

A schematic representation of a planetary gear set is shown in Fig. 2.11.

The tangential speed of the carrier (at the center of the planets, i.e., at a radius intermediate between sun and ring) is the average of the sun and ring speeds. Indicating with the subscripts s, r, and c the sun, ring, and carrier shafts, the following kinematic constraint can be written:

$$\omega_c(N_r + N_s) = (\omega_r N_r + \omega_s N_s), \qquad (2.20)$$

where N_r and N_s are the number of teeth of the ring and sun gear, respectively. The reason for writing this relation in terms of number of teeth instead of radii is that—in a given gear set—the number of teeth N of each gear is directly proportional to the radius of the respective gear.

Introducing the planetary gear ratio $\rho = N_s/N_r$ (the ratio of the number of teeth of sun to the number of teeth of the ring), the kinematic relation (2.20) is written in

Fig. 2.11 Schematic representation of planetary gear set

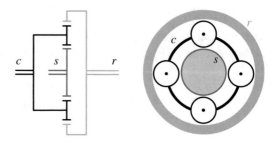

Fig. 2.12 Torque balance on
the planets

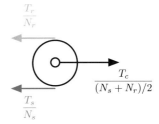

the more compact form:

$$(1 + \rho)\omega_c = \rho\omega_s + \omega_r. \tag{2.21}$$

The torque at the carrier at steady state is equally split between the sun and the ring; for the equilibrium of the planets (Fig. 2.12), the following torque equations hold:

$$\frac{T_c}{(N_r + N_s)} = \frac{T_r}{N_r}, \tag{2.22a}$$

$$\frac{T_c}{(N_r + N_s)} = \frac{T_s}{N_s}, \tag{2.22b}$$

where, again, the number of teeth are used instead of the radii. Using the planetary gear ratio $\rho = N_s/N_r$, the equilibrium equations become:

$$T_c = (1 + \rho)T_r, \tag{2.23a}$$

$$T_s = \rho T_r. \tag{2.23b}$$

Equations (2.21) and (2.23a, 2.23b) are the basis for modeling planetary gear sets. The torque equations (2.23a, 2.23b) are only valid in steady-state conditions and neglect losses, but can be used with reasonable accuracy in vehicle-level models.

2.4.5 *Wheels, Brakes, and Tires*

The wheel represents the link between the powertrain and the external environment. Its model includes the motion of the wheel and the effect of the brakes, calculating the forces at the interface between tire and road surface. The tractive force is calculated given the powertrain torque, the brake signal and the vertical load on the wheel. A quasi-static model is usually sufficient, while dynamic tire models (see, for example, [4]) are typically used in models for vehicle lateral dynamics (handling models).

The static tire model could be defined a *perfect rolling* model, in which the torque applied to the wheel shaft is completely transformed into tractive force considering pure rolling motion between the tire and the road, and neglecting tire deformation. These hypotheses work well for driving in normal conditions (not extreme accelerations) on roads with good adherence (dry asphalt). Low-adherence roads or extreme maneuvers require more accurate tire models to predict vehicle behavior in terms of speed dynamics.

The brakes can be modeled as an additional torque that reduces the net torque acting on the tire. The brake torque is proportional to the brake input signal. Therefore the net tractive force acting on the wheels is

$$F_{trac} = \frac{1}{R_{wh}} \cdot \left(T_{pwt} - T_{brake}\right) \tag{2.24}$$

where T_{pwt} is the torque generated by the powertrain at the wheel shaft, T_{brake} the braking torque, and R_{wh} the wheel radius.

The wheel speed is

$$\omega_{wh} = \frac{v_{veh}}{R_{wh}}, \tag{2.25}$$

being v_{veh} the longitudinal vehicle speed.

The value of longitudinal force is bounded by the vertical load acting on the wheel:

$$- F_z v_{x,max} \leq F_{trac} \leq F_z v_{x,max}, \tag{2.26}$$

where F_z is the vertical force on the wheel, and $v_{x,max}$ is the peak value of the road/tire friction coefficient (usually around 0.8–0.9 for dry asphalt). In order to maintain proper vehicle stability and maximize braking efficiency, the braking action must be distributed between front and rear axles according to the normal load acting on each, also accounting for the longitudinal load transfer generated by the deceleration. From (2.1), the total tractive force during braking is:

$$F_{trac} = M_{veh}\dot{v}_{veh} + F_{roll} + F_{aero} + F_{grade}. \tag{2.27}$$

This should be distributed between the front and rear axle (*f* and *r*) proportionally to the vertical load on each, i.e.:

$$\frac{F_{trac,f}}{F_{trac}} = \frac{b}{a+b} - \frac{M_{veh}\dot{v}_{veh}h_{CG}}{M_{veh}g(a+b)} \tag{2.28}$$

$$\frac{F_{trac,r}}{F_{trac}} = \frac{a}{a+b} + \frac{M_{veh}\dot{v}_{veh}h_{CG}}{M_{veh}g(a+b)} \tag{2.29}$$

where a and b are the distances of the center of gravity (CG) from the front and rear axle respectively, and h_{CG} its height from the ground. The terms that include the acceleration $\dot{v}_{veh}$ represent the dynamic load transfer, from the rear axle to the front axle during deceleration (negative $\dot{v}_{veh}$), and in the opposite direction during acceleration. In most passenger vehicles, the powertrain generates torque only on one of the two axles. In that case, regenerative braking can only be applied to that axle, and must be appropriately balanced by conventional braking on the other axle. From the energy management standpoint, this means that not all the braking torque can be regenerated, but only the fraction of it that is applied at the traction axle, i.e., (2.28) for front-wheel drive or (2.29) for rear-wheel drive vehicles.

2.4.6 Electric Machines

The electric machines can be modeled using an approach similar to the one used for the engine, i.e., based on maps of torque and efficiency. Desired values of electrical power or torque can be used as a control input. Rotor inertia is the main dynamic element that is usually modeled, as the electrical dynamics are very fast in comparison with the inertial dynamics or the engine dynamics.

The relation between torque at the shaft and electric power is provided by an efficiency map, which can be expressed as a function of speed and torque, or speed and electrical power (depending on the implementation).

The efficiency map can also include the power electronics between the main electric bus and the machine to provide directly the electric power exchanged with the battery; otherwise, an explicit power electronics efficiency should be included in the model between the electric machine and the battery.

The efficiency model can be expressed as,

$$P_{mech} = \omega_{em} \cdot T_{em} = \begin{cases} \eta_{em}(\omega_{em}, P_{elec}) \cdot P_{elec} & \text{if } P_{elec} \geq 0 \text{ (motoring mode)}, \\ \frac{1}{\eta_{em}(\omega_{em}, P_{elec})} P_{elec} & \text{if } P_{elec} < 0 \text{ (generating mode)}, \end{cases}$$
$$(2.30)$$

or, if electric power is the desired output, as

$$P_{elec} = \begin{cases} \frac{1}{\eta(\omega_{em}, T_{em})} P_{mech} = \frac{1}{\eta_{em}(\omega_{em}, T_{em})} \omega \cdot T_{em} & \text{if } P_{elec} \geq 0 \text{ (motoring mode)}, \\ \eta_{em}(\omega_{em}, T) \cdot P_{mech} = \eta_{em}(\omega_{em}, T_{em}) \cdot \omega_{em} \cdot T_{em} & \text{if } P_{elec} < 0 \text{ (generating mode)}. \end{cases}$$
$$(2.31)$$

An example of efficiency map for an electric motor is shown in Fig. 2.13.

2.4.7 Batteries

Electrochemical energy storage systems such as batteries and capacitors are key components of hybrid electric vehicles. A variety of models have been proposed to evaluate their interaction with the rest of the powertrain [8].

Fig. 2.13 Example of
electric motor efficiency map
(elaboration of data in [7])

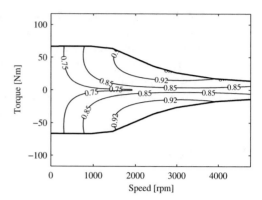

Accurately modeling battery dynamics in hybrid electric vehicles is critical and
not trivial, because the main variables that characterize battery operation, i.e. state
of charge, voltage, current and temperature, are dynamically related to each other in
a highly nonlinear fashion. In general, the objective of the battery model in a vehicle
simulator is to predict the change in state of charge given the electrical load.

The state of charge (SOC) is defined as the amount of electrical charge stored in
the battery, relative to the total charge capacity:

$$SOC(t) = \frac{Q(t)}{Q_{nom}}, \tag{2.32}$$

where Q_{nom} is the nominal charge capacity, and $Q(t)$ the amount of charge currently
stored. The SOC dynamics are given by:

$$S\dot{O}C(t) = \begin{cases} -\frac{1}{\eta_{coul}}\frac{I(t)}{Q_{nom}} & \text{if } I(t) > 0 \\ -\eta_{coul}\frac{I(t)}{Q_{nom}} & \text{if } I(t) < 0 \end{cases} \tag{2.33}$$

where I is the battery current (positive during discharge), η_{coul} is the Coulombic
efficiency [1] or charge efficiency, which accounts for charge losses and depends on
current operating conditions (mainly current intensity and temperature).

Calculating the state of charge (or, better, its variation) by integration of (2.33)
appears to be relatively straightforward, if the capacity is assumed to be a constant,
known parameter. In reality, the battery capacity and coulombic efficiency change
according to several parameters, and the numerical integration is reliable only in
simulation in the absence of measurement error and noise, which makes reliable
state of charge estimation a significant portion of the actual battery management
system (BMS) [8].

In order to correlate the battery current and voltage to the power exchanged with
the rest of the powertrain, a circuit model of the battery can be used.

A simple dynamic model is a circuit like the one in Fig. 2.14, which represents a
second-order approximation.

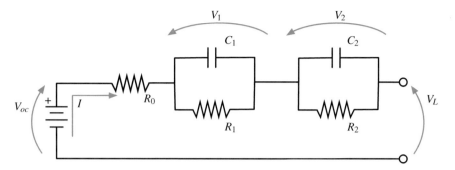

Fig. 2.14 Battery equivalent circuit-based model (second-order)

Fig. 2.15 Battery circuit model (no dynamics)

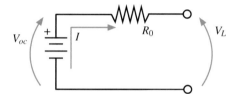

The series resistance R_0 represents the Ohmic losses due to actual resistance of the wires and the electrodes and also to the dissipative phenomena that reduce the net power available at the terminals; the resistances R_1, R_2 and the capacitances C_1, C_2 are used to model the dynamic response of the battery. The values of the parameters are estimated using curve fitting of experimental data, and are generally variable with the operating conditions (temperature, state of charge, current directionality). Other models of the same kind, with more or fewer R–C branches in series, can be used depending on the required model accuracy. However, the number of parameters to be identified increases with the model order. Very often, simpler models without any R–C branch (Fig. 2.15) can also be used if the voltage dynamics can be neglected, for example in quasi-static models focusing exclusively on efficiency considerations. When no detailed data from battery testing is available, circuit models with a single, constant R_0 may be the only option.

The equations of the circuit in Fig. 2.14 are:

$$V_L = V_{oc} - R_0 I - \sum_{i=1}^{n} V_i, \tag{2.34}$$

$$C_i \frac{dV_i}{dt} = I - \frac{V_i}{R_i}, \tag{2.35}$$

where V_L is the load voltage at the battery terminals, V_{oc} is the open circuit voltage, i.e., the voltage of the battery when it is not connected to any load ($I = 0$), R_0 the series resistance, V_i the voltage across the ith R–C branch (characterized by the resistance

R_i and the capacitance C_i), n is the order of the dynamic model considered, i.e., the number of R–C branches. In the example shown, $n = 2$. The capacitance C_i and the resistance R_i can change with the direction (charge or discharge) and amplitude of the current and with other operating conditions, such as temperature and state of charge; the variation can be taken into account by expressing the parameters as maps (tables) instead of constants.

If voltage dynamics are neglected and the battery circuit is represented without R–C branches as in Fig. 2.15, the circuit equation is easily written as a function of the terminal power P_{batt}:

$$P_{batt} = V_L \cdot I = V_{oc}I - R_0 I^2, \tag{2.36}$$

thus providing an explicit expression of the current as a function of power:

$$I = \frac{V_{oc}}{2R_0} - \sqrt{\left(\frac{V_{oc}}{2R_0}\right)^2 - \frac{P_{batt}}{R_0}}. \tag{2.37}$$

The circuit representation of Figs. 2.14 and 2.15 and the corresponding equations are referred to the entire battery pack. This is usually composed by many cells connected in series (strings), and possibly several strings in parallels. The electrical parameters of the circuit models are those of the entire pack, which can be computed from the values of each cell as follows:

$$V_{oc} = N_S V_{oc,cell}, \tag{2.38}$$

$$R_i = \frac{N_S}{N_P} R_{i,cell}, \quad i = 0, \ldots, n \tag{2.39}$$

$$C_i = \frac{N_P}{N_S} C_{i,cell}, \quad i = 1, \ldots, n \tag{2.40}$$

where N_S is the number of cells in series in each string, and N_P is the number of strings in parallel.[4]

The open circuit voltage V_{oc} is a typical characteristic of the battery (or, better, of its cells) and is primarily a function of the state of charge. An example of variation of the open circuit voltage V_{oc} with the state of charge for a single Li-Ion cell is shown in Fig. 2.16. The figure also shows the internal resistance of the same cell. It is common practice to refer to the value of the current in terms of its C-rate, i.e., as a fraction of the battery capacity (expressed in Ah): for example, if the capacity is 6.5 Ah, a current of 1 C corresponds to 6.5 A, 10 C–65 A, 0.1 C–0.65 A. Steady-state characteristics of the battery, such as those of Fig. 2.16, are typically obtained using a current of 1 C.

---

[4]Equations (2.38)–(2.40) are simplifications based on the assumption of ideal cells, all identical. In reality, each cell may have slightly different characteristics, for manufacturing issues and for normal imbalance during operation.

Fig. 2.16 Typical
characteristics of open
circuit voltage and internal
resistance for a Li-Ion cell
(Data referred to one blended
cathode composed of
layered-oxide positive
electrodes and spinel oxide
positive electrodes pouch
Li-Ion cell, obtained from
experiments at the Center for
Automotive Research—The
Ohio State University)

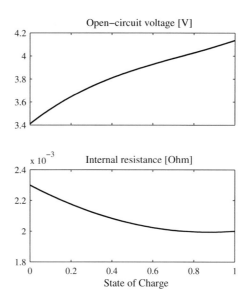

Apparently, integration of the current is sufficient to calculate the state of charge; however, in real-world applications, this is not stable (for numerical drift) nor accurate (for modeling approximations); therefore, more complex *SOC* estimation algorithms are used, which provide an estimate of the battery state of charge given available measurements of terminal voltage and current.

An important issue related to battery usage in hybrid electric vehicles is their aging, due to the aggressive loading cycles to which they are subjected. Battery aging manifests itself as loss of capacity and increase of internal resistance and can reduce vehicle performance; recent and ongoing research is devoted to determine a suitable model that can predict the amount of residual life given the loading cycles [9–13].

The dependence on aging does not affect battery performance in the short term and is only apparent over a long period of time, that exceeds any typical power-train/vehicle simulation horizon; therefore, it is not taken into account as a dynamic effect in this study.

2.4.8 Engine Accessories and Auxiliary Loads

The engine powers several auxiliaries, such as air conditioning, power steering, alternator for small electric loads, etc. A simplified modeling approach is often used, for the lack of detailed data and load cycles for all the components, using the net power as an input in the form of a load cycle, and computing the torque as the ratio of power to speed, including an efficiency computed using a look-up table, curve fitting, or constant value (depending on the case).

Auxiliary loads are important especially in heavy-duty vehicles with specialized use, but may be significant also in passenger vehicles. For example, the power demand of the air-conditioning system in a compact car can be as high as 10 % of the maximum engine power. Due to the great variety of possible auxiliary loads in a vehicle, an attempt to first-principle modeling cannot be generalized and must be derived for the specific application. For this reason, instead of dealing with detailed modeling of the accessories, the usual approach is to assume a known torque or power profile generated by the auxiliary loads at the engine shaft (if they are mechanically driven), or electric power at the bus interface (if they are electrically powered). In many cases, especially for passenger cars, the auxiliary loads are assumed to be constant for the entire driving cycle, using an estimate average value, whose order of magnitude is 1–4 kW.

References

1. L. Guzzella, A. Sciarretta, *Vehicle Propulsion Systems: Introduction to Modeling and Optimization* (Springer, Berlin, 2013)
2. L. Serrao, C. Hubert, G. Rizzoni, Dynamic modeling of heavy-duty hybrid electric vehicles, in *Proceedings of the 2007 ASME International Mechanical Engineering Congress and Exposition* (2007)
3. P. Sendur, J.L. Stein, H. Peng, L. Louca, An algorithm for the selection of physical system model order based on desired state accuracy and computational efficiency, in *Proceedings of the 2003 ASME International Mechanical Engineering Congress and Exposition*. Washington, DC, USA (2003)
4. H.B. Pacejka, *Tire and Vehicle Dynamics* (Butterworth-Heinemann, Oxford, 2012)
5. P. Pisu, C. Cantemir, N. Dembski, G. Rizzoni, L. Serrao, J. Josephson, J. Russell, Evaluation of powertrain solutions for future tactical truck vehicle systems, in *Proceedings of SPIE*, vol. 6228 (2006)
6. A.J. Kotwicki, Dynamic models for torque converter equipped vehicles. SAE Technical Paper No. 820393 (1982)
7. T.A. Burress, S.L. Campbell, C.L. Coomer, C.W. Ayers, A.A. Wereszczak, J.P. Cunningham, L.D. Marlino, L.E. Seiber, H.T. Lin, Evaluation of the 2010 Toyota Prius hybrid synergy drive system. Technical report, Oak Ridge National Laboratory (2011)
8. C.D. Rahn, C.-Y. Wang, *Battery Systems Engineering* (Wiley, Oxford, 2013)
9. S. Drouilhet, B. Johnson, A battery life prediction method for hybrid power applications, in *Proceedings of the 35th AIAA Aerospace Sciences Meeting and Exhibit*, Reno, NV (1997)
10. L. Serrao, Z. Chehab, Y. Guezennec, G. Rizzoni, An aging model of Ni-MH batteries for hybrid electric vehicles, in *Proceedings of the 2005 IEEE Vehicle Power and Propulsion Conference (VPP05)* (2005), pp. 78–85
11. Z. Chehab, L. Serrao, Y. Guezennec, G. Rizzoni, Aging characterization of Nickel—Metal Hydride batteries using electrochemical impedance spectroscopy, in *Proceedings of the 2006 ASME International Mechanical Engineering Congress and Exposition* (2006)
12. M. Dubarry, V. Svoboda, R. Hwu, B. Liaw, Capacity and power fading mechanism identification from a commercial cell evaluation. J. Power Sources **165**(2), 566–572 (2007)
13. M. Dubarry, V. Svoboda, R. Hwu, B. Liaw, Capacity loss in rechargeable lithium cells during cycle life testing: the importance of determining state-of-charge. J. Power Sources **174**(2), 1121–1125 (2007)

Chapter 3
The Energy Management Problem in HEVs

3.1 Introduction

Energy management in hybrid vehicles consists in deciding the amount of power delivered at each instant by the energy sources present in the vehicle while meeting several constraints. In this chapter, the role and main features of the energy management controller are introduced, and the problem of designing such a controller is formalized using optimal control theory.

3.2 Energy Management of Hybrid Electric Vehicles

Controlling an HEV includes essentially two sets of tasks. One is the *low-level* or *component-level* control task, where each powertrain component is controlled by using classical feedback control methods. The second task, referred to as *high-level* or *supervisory control*, is responsible for the optimization of the energy flow on-board of the vehicle while maintaining the battery state of charge within a certain range of operation. This layer of control, called Energy Management System (EMS), receives and processes information from the vehicle (ω_{eng}, ω_{gb}, ω_{mot}) and the driver (v_{veh}, a_{veh}, δ) to output the optimal set-points sent to the actuators and executed by the low-level control layer. The EMS also selects the best modes of operations of the hybrid powertrain, including start–stop, power split, and electric launch. The two-task based control scheme of an HEV is shown in Fig. 3.1.

Realistic figures of achievable improvement in fuel economy in HEVs range from 10 % for mild hybrids to more than 30 % for full hybrid vehicles [1]. This potential can be realized only with a sophisticated control system that optimizes energy flow within the vehicle. It has been recognized that the adoption of systematic model-based optimization methods using meaningful objective functions to improve the energy management controllers is the pathway to go in order to achieve near-optimal results in designing the vehicle EMS.

This books presents several model-based optimization techniques for energy management of hybrid electric vehicles.

© The Author(s) 2016   31
S. Onori et al., *Hybrid Electric Vehicles*, SpringerBriefs in Control,
Automation and Robotics, DOI 10.1007/978-1-4471-6781-5_3

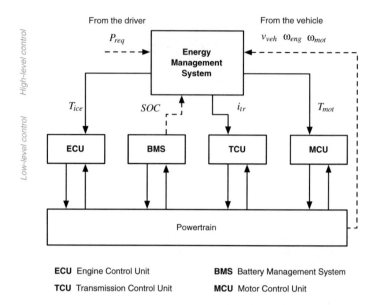

Fig. 3.1 Two-layer control architecture in a hybrid vehicle. The EMS elaborates information from the driving mission and the powertrain components to generate actuator set-points corresponding to the optimal power split between the primary and secondary energy sources (high-level control). The powertrain components control (lower level control) is then performed on single components using traditional closed-loop control methods

In a conventional (non-hybrid) vehicle, there is no need for an energy management strategy: the driver decides the instant power delivery using the brake and accelerator pedals, and, in manual transmission vehicles, decides what gear is engaged at each time. The driver's desires are translated into actions by the low-level control: for example, the engine control unit (ECU) determines the amount of fuel to be injected given the desired torque request; the automatic transmission controller in the TCU decides when to shift gear based on engine conditions and vehicle speed, etc.

In a hybrid vehicle, on the other hand, there is an additional decision that must be taken: how much power is delivered by each of the energy sources on-board of the vehicle. This is why all hybrid vehicles include an energy management controller, interposed between the driver and the component controllers. As mentioned, the aim of the energy management system is to determine the optimal power split between the on-board energy sources. The decision on what to consider optimal depends on the specific application: in most cases, the strategies tend to minimize the fuel consumption, but optimization objectives could also include the minimization of pollutant emissions, maximization of battery life, or—in general—a compromise among all the above goals.

The role of the energy management system in a hybrid vehicle can also be represented as in Fig. 3.2. The outer layer in the figure is the speed control, which is the human driver in a real vehicle and a driver model (typically a PI controller) in

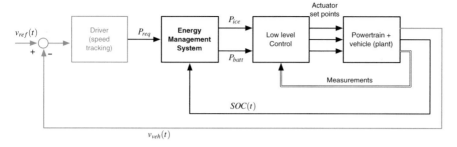

Fig. 3.2 The role of energy management system

simulation. The speed controller decides the total power request P_{req} that the power-train must deliver in order to follow the prescribed velocity profile. The inner layer is the energy management system, which decides how to split the total power request between the on-board energy sources: the rechargeable energy storage system and the internal combustion engine. When designing the energy management strategy, the separation of the two controllers allows to consider only the battery state of charge dynamics as the system state and neglect the vehicle speed, since this is controlled directly by the driver.

3.3   Classification of Energy Management Strategies

Several families of energy management strategies have been proposed in literature. Two general trends can be identified that deal with the energy management problem, namely rule-based and model-based optimization methods [2–4].

The main characteristic of rule-based approaches is their effectiveness in real-time implementation. They do not involve explicit minimization or optimization, but rely on a set of rules to decide the value of the control to apply at each time. Rules are generally designed based on heuristics [5], intuition, or from the knowledge of optimal global solution generated with mathematical models through optimization algorithms [6–8].

In model-based optimization strategies, the optimal actuator set-points are calculated by minimization of a cost function over a fixed and known driving cycle, leading to a global optimal solution. This generates a noncausal solution in that it finds the minimum value of the cost function using knowledge of the future driving information. Although model-based optimization control methods cannot be used directly for real-time implementation and do not directly lend themselves to practical implementation, due to both their preview nature and computational complexity, they constitute a valuable design tool. In fact, they can be used to design rules for online implementation or used as a benchmark solution to evaluate the performance of other control strategies. We can divide model-based optimization methods into numerical and analytical approaches. In numerical optimization methods, like dynamic programming [6, 9], simulated annealing [10], genetic algorithms [11], and stochastic dynamic programming [12, 13], the entire driving cycle is taken into consideration and the global optimum is found numerically.

Analytical optimization methods, on the other hand, use an analytical problem formulation to find the solution in closed, analytical form, or at least provide an analytical formulation that makes the numerical solution faster than the purely numerical methods. Among these methods, Pontryagin's minimum principle [14] is the most important. Equivalent consumption minimization strategy also belongs to this category in that it consists in the minimization, at each time step of the optimization horizon, of an appropriately defined instantaneous cost function. This leads (ideally) to the minimization of the global cost function, if the instantaneous cost function (similar to an instantaneous *equivalent* fuel consumption) is suitably defined. Other model-based strategies consider information about future driving conditions in addition to past and present, for example, using a receding-horizon optimization approach [15–17].

3.4   The Optimal Control Problem in Hybrid Electric Vehicles

Regardless of the powertrain topology, the essence of the HEV control problem is the instantaneous management of the power flows from energy converters to achieve the control objectives. One important characteristic of this problem is that the control objectives are mostly integral in nature (for instance, fuel consumption) or semi-local in time, such as drivability, while the control actions are local in time. Furthermore, the control objectives are often subject to integral constraints, such as maintaining the battery *SOC* within a prescribed range [18]. In general, the energy management problem in a hybrid vehicle can be cast into an optimization problem over a finite time horizon, whose solution can be found in the pool of *optimal control theory* methods which are aimed at finding a control law for a given system such that a certain optimality criterion, usually defined as an integral performance index over a certain time frame, is achieved.

Traditional optimal control techniques can be used only with simple mathematical models of the system, assuming perfect knowledge of the entire optimization horizon (time frame over which the optimization is defined); since both these conditions are usually not respected by real systems, optimal control implementation in a physical dynamic system, whose future is unknown, is necessarily *suboptimal*.

3.4.1 Problem Formulation

In this section, we formulate the supervisory control problem when the total mass of fuel, m_f, [g], is being minimized during a driving mission.[1]

The optimal energy management problem in a hybrid electric vehicle consists of finding the control $u(t)$ that leads to the minimization of the fuel consumed, m_f, over a trip of length t_f (starting at $t_0 = 0$). This is equivalent to minimizing the integral performance index J:

$$J = \int_{t_0}^{t_f} \dot{m}_f(u(t), t)dt, \tag{3.1}$$

where $\dot{m}_f$ [g/sec] is the mass flow rate of fuel used. The minimization of J is subject to constraints related to physical limitations of the actuators, limitation in the energy stored in the RESS and the requirement to maintain the battery SOC within prescribed limits. This makes the optimal energy management problem a constrained, finite-time optimal control problem, where the objective function (3.1) is minimized under a set of both local and global constraints on the state and control variables, as outlined in the following.

System Dynamics. From (2.33) and with $x(t) = SOC$ and $u(t) = P_{batt}$, the system dynamics are given by:

$$\dot{x}(t) = -\frac{1}{\eta_{coul}^{\text{sign}(I(t))}} \frac{I(t)}{Q_{nom}}. \tag{3.2}$$

For the control design, a control-oriented model of the battery is used based on a zero-th order equivalent circuit model, shown in Fig. 3.3, whose parameters are: the equivalent resistance, $R_0(SOC)$ and the open circuit voltage, $V_{oc}(SOC)$.[2] The SOC variation can be expressed as a function of the battery power:

$$V_L(t)I(t) = P_{batt}(t) = u(t) = V_{oc}(x)I(t) - R_0(x)I^2(x) \tag{3.3}$$

Solving (3.3) for the current and replacing it into (3.2) yields:

$$\dot{x} = -\frac{1}{\eta_{coul}^{\text{sign}(I(t))} Q_{nom}} \left[\frac{V_{oc}(x)}{2R_0(x)} - \sqrt{\left(\frac{V_{oc}(x)}{2R_0(x)}\right)^2 - \frac{u(t)}{R_0(x)}} \right] \tag{3.4}$$

---

[1] We limit our focus to the problem of fuel consumption minimization, with no inclusion of drivability considerations. Typically, the gear shifting optimization pertains to the transmission control and it is not an objective of the supervisory control. The optimization of the gear shifting strategy would require the formulation of an optimal control problem which includes both continuous time and discrete time dynamics. In this work, we assume that the transmission controller operates independently of the supervisory controller, therefore the gear shifting strategy is treated as a known external input to the energy management system.

[2] In general, both the resistance and open circuit voltage are dependent also on temperature, θ. In this study we neglect the effect of the temperature on the battery parameters.

Fig. 3.3 Zero-th order
equivalent circuit-based
model of the battery used for
control design

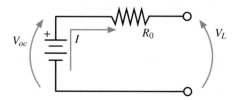

which is written in the general form:

$$\dot{x}(t) = f(x(t), u(t)). \tag{3.5}$$

The optimization problem is subject to several constraints. Some of them are integral in nature (for example, the final SOC target); some are local (instantaneous power limits, state of charge boundaries).

Global Constraints. The final SOC value $x(t_f)$ should reach a predefined value x_{target}:

$$x(t_f) = x_{target}. \tag{3.6}$$

In other words, it is required that

$$x(t_f) - x_{target} = \Delta x = 0. \tag{3.7}$$

In a charge-sustaining HEV, the net energy from the battery should be zero over a given driving mission, meaning that the SOC at the end of the driving cycle, $x(t_f)$, should be the same as that at the beginning of the driving cycle, i.e., $x_{target} = x(t_0)$. In other applications, for example PHEVs, where a positive net energy consumption is desired (battery depletion), the SOC target value may be lower than initial one. Equation (3.7) defines the global state constraints of the control problem, and is justified mainly as a way to compare the results of different solutions by guaranteeing that they start and reach the same level of battery energy. In practical vehicle applications, it is sufficient to keep the SOC between two boundary values, in that some difference between the desired and actual SOC at the end of a cycle is acceptable and does not affect the vehicle functionality.

Local Constraints. Local constraints are imposed on the state and control variables. The local (or instantaneous) constraints on the state concern the fact that the state of charge must remain between a maximum and a minimum value (to make the battery work at high efficiency and preserve its cycle life), whereas local constraints on the control variables are imposed to guarantee the physical operation limits (maximum and minimum engine, motor and generator torque and speed, and battery power).

Overall, the local constraints are:

$$\begin{aligned}
SOC_{min} &\le SOC(t) \le SOC_{max}, \\
P_{batt,min} &\le P_{batt}(t) \le P_{batt,max}, \\
T_{x,min} &\le T_x(t) \le T_{x,max}, \\
\omega_{x,min} &\le \omega_x(t) \le \omega_{x,max}, \quad x = eng,\ mot,\ gen,
\end{aligned} \tag{3.8}$$

where the last two inequalities in (3.8) represent limitations on the instantaneous engine and motor torque and speed, respectively; $(\cdot)_{min}$, $(\cdot)_{max}$ are the minimum and maximum value of SOC/power/torque/speed at each instant.

Other local constraints could be included related to drivability or comfort issues: for example, it is usually a good idea to limit the frequency of switching between operating modes. In this book, we limit our discussion only to local constraints expressed by (3.8).

In addition to meeting the local and global constraints, at each instant the supervisory controller ensures that the total power demand at the wheels is satisfied.

Problem 3.1 The **optimal energy management problem** in a charge-sustaining HEV consists in finding the control sequence u^* that minimizes the cost function (3.1) while meeting the dynamic state constraint (3.5), the global state constraint (3.7) and local state and control constraints (3.8).

Note
The constraints on the state of charge are very important for its control. A *charge-sustaining* hybrid vehicle is so called because its battery SOC at the beginning and the end of each trip is nominally the same, which means that the vehicle maintains its own electrical charge without need for external recharging. On the other hand, a *charge-depleting* or *plug-in* hybrid vehicle (PHEV) can be recharged using an electric outlet, and therefore the SOC after a trip can be lower than it was at the beginning. In charge-sustaining HEVs ultimately all the energy consumed derives from fuel, while in PHEVs part of can be obtained from the electric power grid.

3.4.2　General Problem Formulation

The energy management problem as defined in Problem 3.1 is a particular case of a more general optimal control problem in case fuel consumption is the objective to minimize and charge-sustaining operation is enforced through hard constraints (3.7). Nonetheless, energy management problems for different applications can be formulated from the general formulation reported below.

Consider a generic dynamic system with state equation

$$\dot{\mathbf{x}}(t) = f\left(\mathbf{x}(t), \mathbf{u}(t), t\right), \tag{3.9}$$

where $\mathbf{x}(t) \in \mathbb{R}^n$ indicates the state vector, and $\mathbf{u}(t) \in \mathbb{R}^p$ is the control vector.

The charge-sustaining constraint has been taken into account as a hard constraint in (3.7) by requiring that the energy stored at the end of the mission equal the value at the beginning of the mission. Alternatively, this constraint can be enforced as a soft constraint, that is, by penalizing deviations from the initial value of the energy stored at the end of the mission.

This is done by means of a penalty function $\phi(\mathbf{x}(t_f))$ (generally, a function of the difference $\mathbf{x}(t_f) - \mathbf{x}(t_0)$) to be added to the performance index (3.1) to obtain a charge-sustaining performance index of the form [2]:

$$J = \phi(\mathbf{x}(t_f)) + \int_{t_0}^{t_f} \dot{m}_f(\mathbf{u}(t), t)\, dt. \tag{3.10}$$

Soft constraints modify the cost function with the term $\phi(\mathbf{x}(t_f))$ in order to induce the final value of the constrained variable to be close, but not necessarily identical, to the desired target. For instance, in [19] the soft constraint is represented by a quadratic function of the difference $\mathbf{x}(t_f) - \mathbf{x}(t_0)$, namely $\phi = \alpha||\mathbf{x}(t_f) - \mathbf{x}(t_0)||_2^2$ where α is a positive weighting factor.

Using a quadratic function tends to equally penalize positive and negative deviations from the target SOC, x_{target}, regardless of the sign of the deviation. In [2], a linear penalty function of the type

$$\phi(\mathbf{x}(t_f)) = w(\mathbf{x}(t_0) - \mathbf{x}(t_f)), \tag{3.11}$$

is proposed, where w is a positive constant. With (3.11) the battery use is penalized while favoring the energy stored. The soft constraints (3.11) are in effect a penalty term that can be expressed as:

$$\phi(\mathbf{x}(t_f)) = w \int_0^{t_f} \dot{\mathbf{x}}(t)\, dt. \tag{3.12}$$

Engine exhaust emissions [19–22], battery aging [23], drivability [24, 25], thermal dynamics [26] considerations can be included in the performance index, J, by considering a more general expression

$$J = \phi(\mathbf{x}(t_f)) + \int_{t_0}^{t_f} L(\mathbf{x}(t), \mathbf{u}(t), t)\, dt \tag{3.13}$$

where $L(\cdot)$ is the cost function. Several objectives can be combined together by introducing a weighting factor for each [27], for instance: fuel consumption and battery aging.

The optimal control problem in the time interval $t \in [t_0, \; t_f]$ corresponds to choose the law $\mathbf{u} : [t_0, \; t_f] \mapsto \mathbb{R}^p$ that leads to the minimization of the *cost function* (3.13) under dynamic, local, and global constraints.

In general, the constraints on the states can be expressed by defining the set of admissible states as those for which the conditions $\mathbf{G}(\mathbf{x}, t) \leq 0$ are satisfied, i.e.,

$$\Omega_{\mathbf{x}}(t) = \{\mathbf{x} \in \mathbb{R}^n | \mathbf{G}(\mathbf{x}, t) \leq 0\},$$

where the function $\mathbf{G}(\mathbf{x}, t) : \mathbb{R}^n \mapsto \mathbb{R}^m$ represents a set of m inequalities that the components of the state vector must satisfy. For example, if $\mathbf{x}(t)$ must remain between $\mathbf{x}_{min}$ and $\mathbf{x}_{max}$, two inequalities can be written (i.e., $m = 2$):

$$G_1(\mathbf{x}(\mathbf{t})) = \mathbf{x}(t) - \mathbf{x}_{max} \leq 0 \qquad (3.14)$$
$$G_2(\mathbf{x}(\mathbf{t})) = \mathbf{x}_{min} - \mathbf{x}(t) \leq 0 \qquad (3.15)$$

Generally, the set of admissible states and controls are defined as:

$$\begin{cases} \mathbf{G}(\mathbf{x}(t)) \leq 0 \\ \mathbf{u}(t) \in U(t) \end{cases} \quad \forall t \in [t_0, \ t_f] \qquad (3.16)$$

where $U(t)$ indicates the set of admissible control values at time t.

The local constraints (3.16) are instantaneous conditions that must be satisfied at each instant of time. The notation $\mathbf{G}(\mathbf{x}(t), t) \leq 0$ is generic and the function $\mathbf{G}(\cdot)$ is, in general vectorial. In most cases, it has two components representing the two inequalities $\mathbf{x}_{min}(t) \leq \mathbf{x}(t) \leq \mathbf{x}_{max}(t)$.

The general optimal control problem can be stated as follows:

Problem 3.2 The **constrained-finite time horizon optimal control problem** consists in finding the control vector $\mathbf{u}^*$ that minimizes the cost function (3.13) while meeting the dynamic state constraints (3.9), and the local state and control constraints (3.16).

In the following chapters, techniques and methods to solve Problem 3.1 are presented and discussed in depth.

References

1. L. Guzzella, A. Sciarretta, *Vehicle Propulsion Systems: Introduction to Modeling and Optimization* (Springer, Berlin, 2013)
2. A. Sciarretta, L. Guzzella, Control of hybrid electric vehicles. IEEE Control Syst. Mag. **27**(2), 60–70 (2007)
3. F. Salmasi, Control strategies for hybrid electric vehicles: evolution, classification, comparison, and future trends. IEEE Trans. Veh. Technol. **56**(5), 2393–2404 (2007)
4. S. Onori, in *Model-Based Optimal Energy Management Strategies for Hybrid Electric Vehicles*, ed. by H. Waschl, I. Kolmanovsky, M. Steinbuch, L. del Re. Lecture Notes in Control and Information Sciences, vol. 455 (Springer, New York, 2014), pp. 199–218
5. B. Baumann, G. Washington, B. Glenn, G. Rizzoni, Mechatronic design and control of hybrid electric vehicles. IEEE/ASME Trans. Mechatron. **5**(1), 58–72 (2000)
6. C. Lin, J. Kang, J. Grizzle, H. Peng, Energy management strategy for a parallel hybrid electric truck, in *Proceedings of the 2001 American Control Conference*, vol. 4 (2001), pp. 2878–2883

7. D. Bianchi, L. Rolando, L. Serrao, S. Onori, G. Rizzoni, N. Al-Khayat, T.M. Hsieh, P. Kang, Layered control strategies for hybrid electric vehicles based on optimal control. Int. J. Electr. Hybrid Veh. **3**(2), 191–217 (2011)
8. R. Biasini, S. Onori, G. Rizzoni, A rule-based energy management strategy for hybrid medium duty truck. Int. J. Powertrains **2**(2/3), 232–261 (2013)
9. D. Bertsekas, *Dynamic Programming and Optimal Control* (Athena Scientific, Belmont, 1995)
10. S. Delprat, G. Paganelli, T.M. Guerra, J.J. Santin, M. Delhorn, E. Combes, Algorithmic optimization tool for the evaluation of HEV control strategies, in *Proceedings of EVS 99* (1999)
11. A. Piccolo, L. Ippolito, V. Galdi, A. Vaccaro, Optimization of energy flow management in hybrid electric vehicles via genetic algorithms, in *Proceedings of the 2001 IEEE/ASME International Conference on Advanced Intelligent Mechatronics* (2001)
12. L. Kolmanovsky, I. Siverguina, B. Lygoe, Optimization of powertrain operating policy for feasibility assessment and calibration: stochastic dynamic programming approach, in *Proceedings of the 2002 American Control Conference* (2002), pp. 1425–1430
13. S.J. Moura, H.K. Fathy, D.S. Callaway, J.L. Stein, A stochastic optimal control approach for power management in plug-in hybrid electric vehicles. IEEE Trans. Control Syst. Technol. **19**(6), 545–555 (2011)
14. H.P. Geering, *Optimal Control with Engineering Applications* (Springer, Berlin, 2007)
15. E. Nuijten, M. Koot, J. Kessels, B. de Jager, M. Heemels, W. Hendrix, P. van den Bosch, Advanced energy management strategies for vehicle power nets, in *Proceedings of EAEC 9th International Congress: European Automotive Industry Driving Global Changes* (2003)
16. M. Salman, M. Chang, J. Chen, Predictive energy management strategies for hybrid vehicles, in *Proceedings of the 2005 IEEE Vehicle Power and Propulsion Conference* (2005)
17. H. Borhan, A. Vahidi, A.M. Phillips, M.L. Kuang, I. Kolmanovsky, S.D. Cairano, MPC-based energy management of a power-split hybrid electric vehicle. IEEE Trans. Control Syst. Technol. **19**, 1–11 (2011)
18. G. Rizzoni, H. Peng, Hybrid and electrified vehicles: the role of dynamics and control. *Mechanical Engineering-CIME* (American Society of Mechanical Engineers, New York, 2013)
19. C. Lin, H. Peng, J. Grizzle, J. Kang, Power management strategy for a parallel hybrid electric truck. IEEE Trans. Control Syst. Technol. **11**(6), 839–849 (2003)
20. S. Delprat, J. Lauber, T. Guerra, J. Rimaux, Control of a parallel hybrid powertrain: optimal control. IEEE Trans. Veh. Technol. **53**(3), 872–881 (2004)
21. V.H. Johnson, K.B. Wipke, D.J. Rausen, HEV control strategy for real-time optimization of fuel economy and emissions. SAE paper 2000-01-1543 (2000)
22. K. Kelly, M. Mihalic, M. Zolot, Battery usage and thermal performance of the Toyota Prius and Honda Insight during chassis dynamometer testing, in *The Seventeenth Annual Battery Conference on Applications and Advances*, vol. 2002 (2002), pp. 247–252
23. L. Serrao, S. Onori, A. Sciarretta, Y. Guezennec, G. Rizzoni, Optimal energy management of hybrid electric vehicles including battery aging, in *Proceedings of the 2011 American Control Conference (ACC)* (2011), pp. 2125–2130
24. P. Pisu, K. Koprubasi, G. Rizzoni, Energy management and drivability control problems for hybrid electric vehicles, in *44th IEEE Conference on Decision and Control and 2005 European Control Conference, CDC-ECC'05* (2005), pp. 1824–1830
25. D. Opila, X. Wang, R. McGee, R. Gillespie, J. Cook, J. Grizzle, An energy management controller to optimally trade off fuel economy and drivability for hybrid vehicles. IEEE Trans. Control Syst. Technol. **20**(6), 1490–1505 (2012)
26. F. Merz, A. Sciarretta, J.-C. Dabadie, L. Serrao, On the optimal thermal management of hybrid-electric vehicles with heat recovery systems. Oil Gas Sci. Technol.—Rev. IFP Energ. Nouv. **67**(4), 601–612 (2012)
27. L. Serrao, A. Sciarretta, O. Grondin, A. Chasse, Y. Creff, D.D. Domenico, P. Pognant-Gros, C. Querel, L. Thibault, Open issues in supervisory control of hybrid electric vehicles: a unified approach using optimal control methods. Oil Gas Sci. Technol.—Rev. IFP Energ. Nouv. **68**, 137–147 (2013)

Chapter 4
Dynamic Programming

4.1 Introduction

In this chapter, we present the first of a series of methods discussed in this brief to solve the optimal control problem formulated in Chap. 3. The dynamic programming is a numerical method that finds the global optimal solution by operating backwards in time.

> *Life can only be understood backwards; but it must be lived forwards.*
>
> (S. Kierkegaard)

4.2 General Formulation

Dynamic programming is a numerical method for solving multistage decision-making problems [1, 2]. It is capable of providing the optimal solution to problems of any complexity level (in the limits of computational capabilities); however, it is noncausal and is only implementable in simulation environment, because it requires a priori information about the entire optimization horizon. The study of Dynamic programming dates back to Richard Bellman, who wrote the first book on the subject in 1957 [2], where he stated what is today called *Bellman's principle of optimality*:

> An optimal policy has the property that whatever the initial state and initial decision are, the remaining decisions must constitute an optimal policy with regard to the state resulting from the first decision.

In other words, from any point on an optimal trajectory the remaining trajectory is optimal for the corresponding problem initiated at that point.

Consider the discrete-time system

$$x_{k+1} = f_k(x_k, u_k)$$

where k takes integer values, say $k = 0, 1, \ldots$. Let u_k be the control variable whose value is to be chosen at time k. Both the state x and the control u are bounded

© The Author(s) 2016
S. Onori et al., *Hybrid Electric Vehicles*, SpringerBriefs in Control,
Automation and Robotics, DOI 10.1007/978-1-4471-6781-5_4

and discretized, i.e., they can assume values in their respective domains: $u_k \in U_k$ and $x_k \in \Omega_k$.

Let us consider the following control policy over the first N time steps:

$$u = \{u_0, u_1, \ldots, u_{N-1}\}$$

The cost of the policy u, starting at initial conditions x_0, is

$$J(x_0, u) = L_N(x_N) + \sum_{k=1}^{N-1} L_k(x_k, u_k) \tag{4.1}$$

where L_k is the instantaneous cost function (the same as the integrand in the continuous-time formulation (3.13)), also called *arc cost* in the context of dynamic programming. The cost function obtained with the optimal solution is

$$J^*(x_0) = \min_u J(x_0, u) \tag{4.2}$$

and the corresponding optimal policy is $u^* = \{u_1^*, u_2^*, \ldots, u_{N-1}^*\}$.

Consider now the "tail subproblem" of minimizing the *cost-to-go Y* from time i (and state x_i) to time N:

$$Y(x_i, i) = L_N(x_N) + \sum_{k=i}^{N-1} L_k(x_k, u_k), \tag{4.3}$$

which corresponds to the last part of the overall problem. Bellman's principle of optimality states that the "tail policy" $\{u_i^*, u_{i+1}^*, \ldots, u_{N-1}^*\}$ is the optimal policy for the tail subproblem.[1] This statement finds an analytical justification in the induction principle [1].

The dynamic programming algorithm is based on Bellman's principle of optimality. Starting from the final step N, the algorithm proceeds backward using the sequence of controls that generate the optimal cost-to-go, i.e.,

$$u_k = \mu^*(x_k, k) = \arg \min_{u \in U_k} \left(L_k(x_k, u) + Y_{k+1}\big(f_k(x_k, u_k), u_k\big) \right) \tag{4.4}$$

for $k = N - 1, N - 2, \ldots, 1$.

$Y(x_1, 1)$, generated at the last iteration, is equal to the optimal (minimum) cost $J^*(x_0)$. $Y(x_N, N) = L_N(x_N)$ is the terminal cost, which depends on the final state x_N. Note that the symbol $Y(x_k, k)$ denotes the *optimal* cost-to-go from state x_k (at time k) to the end of the optimization horizon, while $Y_k(x_k, u_k)$ is a function that depends on the control value u_k, and represents the alternative values that the

---

[1] In the tail subproblem you are at x_i at time i and wish to minimize the cost-to-go from time i to N.

cost-to-go from that same state can assume, depending on the control u_k. In other words, $Y(x_k, k)$ is the minimum value that $Y_k(x_k, u_k)$ can assume as u_k changes. The optimal control sequence can be found by proceeding backward from the final state, choosing at each step the control that minimizes the cost-to-go $Y_k(x_k, u_k)$, and storing in a matrix μ^* the optimal choice at each time instant k and state value x_k.

Since the state values are discretized in the algorithm, but most physical systems are defined by a continuous state, the application of a given control action within the discrete control set might result in the system reaching a state which is not one of the discretized values Ω_k, but intermediate between two of them. In this case, the computation of the cost at the grid values is based on interpolation.

To summarize [3]:

- The optimal control sequence μ^* is a function only of x_k and k.
- The optimal control law is expressed in closed loop form. It is optimal regardless of the past control policy.
- The Bellman equation is solved by backwards induction: the later policy is decided first.

4.3 Application of DP to the Energy Management Problem in HEVs

Dynamic programming can be used to solve the optimal energy management problem defined in Sect. 3.4.1. The sequence of controls u_k (decisions) represents the power split between the internal combustion engine and the rechargeable energy storage system at successive time steps. The cost corresponds to fuel consumption, energy consumption, emissions, or any other design objective. The set of choices at each instant (set U_k) is determined by considering the state of each powertrain component and the total power request. The number of solution candidates that can be considered and evaluated is a compromise between the computational capabilities and the accuracy of the result: in fact, the minimum cost may not exactly coincide with one of the selected points, but the closer these are to each other, the better the approximation of the optimal solution.

Once the grid of possible power splits, or solution candidates, is created, the procedure outlined earlier can be used, associating a cost to each of the solution candidates. Proceeding backwards (i.e., from the end of the driving cycle), the optimal cost-to-go is calculated for each grid point, and stored in a matrix of costs. When the entire cycle has been examined, the path with the lowest total cost represents the optimal solution.

As an example, consider the case of a series HEV in which the decision variable is the battery power, P_{batt}, which is chosen from a set of admissible values (between a maximum and minimum bound).

Starting from the end of the driving cycle (supposed known), the arc costs of all feasible paths are calculated at each time step; feasible paths are defined as those corresponding to an SOC variation compatible with the limits on the battery power. The cost is computed from the powertrain model as the fuel consumption of the engine corresponding to a given battery power decision, which also corresponds to a given SOC variation.

The objective of the dynamic programming algorithm is to select the optimal sequence of battery power such that the total cost is minimized. Selecting a sequence of battery power values leads to deciding a sequence of values of battery SOC, in that the variation of SOC between time steps is proportional to the integral of the battery power between those steps. The correspondence between the control variable and state variation is univocal.

The selection of battery power and SOC values must satisfy the constraints on the admissible maximum and minimum battery power and state of charge, since only the admissible values are considered. Also, the initial and final values of SOC are set with no effort (state global constraints).

The flowchart in Fig. 4.1 illustrates the implementation of the DP algorithm in its basic form.

The state vector x_{vec} is defined and discretized with a step of δx from a minimum value of x_{min} to a maximum value of x_{max}. The length of the vector x_{vec} is N_x. The same is done for the control vector, u_{vec}, which contains N_u components from the minimum value, u_{min} to a maximum value, u_{max}. At time step $k = N$, the final cost-to-go corresponds to the terminal cost L_N. This is computed for each admissible state value within x_{vec}. If the final state is constrained to a specific value or range of values, only a subset of x_{vec} is admissible at the final time step, meaning that the cost associated to reaching one of the nonadmissible state values is set to infinite. Then, as the time progresses backwards, the arc cost $L_k(x_k, u_k)$ for all the combinations of state values and controls is computed and stored in the matrix $L_k(m, n)$, where the indices m and n correspond to the state and control respectively. The matrix L_k contains the costs of moving from each admissible nodes (each element of x_{vec}) at time k to all the reachable nodes at time $k + 1$ (i.e., the states that are obtained by applying each element of u_{vec}).

Then, the cost-to-go candidates $Y_k(m, n)$, which represent the cost to reach the end of the horizon starting at time k from the state $x_{vec}(m)$ and choosing $u_{vec}(n)$ as the first control action, are computed. The optimal cost-to-go $Y(m, k)$ is obtained by choosing the control u_k that generates the minimum of $Y_k(m, n)$. For each admissible state index m, the control index that generates the optimal cost-to-go at time k is stored in the matrix $\mu^*(m, k)$. This is repeated until the initial time is reached ($k = 1$), at which point the optimal control sequence is found. With the optimal control matrix just computed, $\mu^*(m, k)$, the cycle is then run forward in time and the optimal control and state sequence are computed.

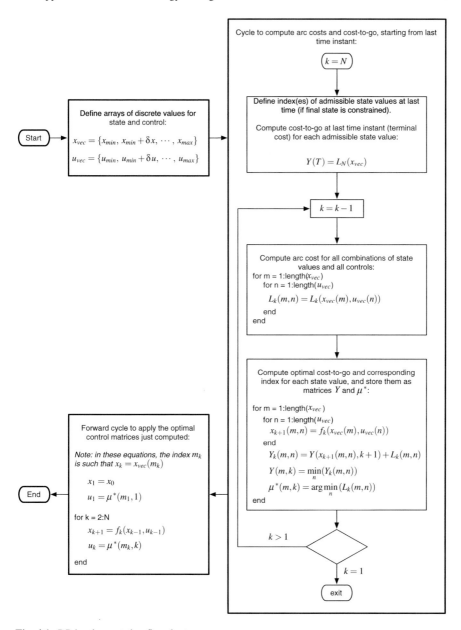

Fig. 4.1 DP implementation flowchart

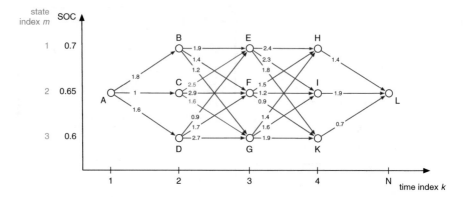

Fig. 4.2 Arc costs in the simplified example. The *points* represent the state values; the *arrows* out of each point represent the control choices (u_{vec}), and point to the state value that would be obtained by applying that control. The numbers on each segment indicate the arc costs $L_k(x, u)$, i.e., the cost incurred (during a single time step) in choosing that control action

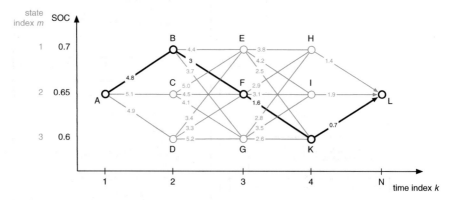

Fig. 4.3 Cost-to-go. The number on each segment starting at a state/time *point* represents the minimum cost-to-go $Y(x, k)$, from that point to the end, that would be incurred by choosing the corresponding control at that time step (and the relative optimal sequence afterwards)

4.3.1   Implementation Example

To illustrate the process, consider the simple example shown in Figs. 4.2 and 4.3. The battery state of charge SOC can take three values: $x_{vec} = \{0.7, 0.65, 0.6\}$, (i.e., $m = 1, 2, 3$) at each time step ($k = 1, \ldots, N$, where $N = 5$). The control (battery power) can take five values: $u_{vec} = \{-0.1, -0.05, 0, 0.05, 0.1\}$: the numerical values indicate the corresponding SOC variation, the index n thus ranges between 1 and 5.

The constraints on the battery power are expressed in terms of maximum and minimum variation of SOC between two subsequent time steps, and depend on the *SOC* value. For example, when the *SOC* is at its maximum value, it cannot increase anymore, and therefore only zero or negative values of u are admissible, i.e., $n = 1, 2, 3$. The arc costs $L_k(x_k, u_k)$ are first computed. These are the costs of moving from all admissible nodes at time k to all the admissible nodes at time $k+1$.

Figure 4.2 shows all the admissible arc costs in this case: for example, at time $k = N - 1 = 4$, all three values of *SOC* are admissible (nodes H, I, K), but only one is accepted at the final time (node L); thus, three arc costs must be defined: H→L ($u = -0.05$), I→L ($u = 0$) and K→L ($u = +0.05$). At time $k = 3$, instead, there are nine possible combinations (from any of the nodes E, F, G to any of the nodes H, I, K). Similar considerations can be made for all other time steps. The arc costs are shown in Fig. 4.2. Once all the arc costs have been determined, the cost-to-go can be calculated, starting from the final point and going backwards (Fig. 4.3). At time $k = 4$, the cost-to-go $Y_4(x_4, u_4)$ of each node H, I, K corresponds to the arc cost because the following time instant is the end of the optimization horizon. At time $k = 3$, the optimal cost-to-go of each node corresponds to the minimum cost associated with moving from that node to the end. So, for node E, the cost-to-go is the one corresponding to the path with minimum cost among the possible alternatives: E→H→L, E→I→L, and E→K→L. The respective values of cost-to-go, computed using the arc costs in Fig. 4.2, are:

$$Y_3(m = 1, n = 3) = 2.4 + 1.4 = 3.8$$
$$Y_3(m = 1, n = 2) = 2.3 + 1.9 = 4.2$$
$$Y_3(m = 1, n = 1) = 1.8 + 0.7 = 2.5.$$

The control indexes $n = 4$ and $n = 5$ are not admissible, since they would overcharge the battery, and for this they are not considered. On the other hand, in node F, the admissible control values are $n = 2, 3, 4$ and in node G they are $n = 3, 4, 5$; the corresponding cost-to-go values are:

$$Y_3(m = 2, n = 4) = 1.5 + 1.4 = 2.9$$
$$Y_3(m = 2, n = 3) = 1.2 + 1.9 = 3.1$$
$$Y_3(m = 2, n = 2) = 0.9 + 0.7 = 1.6.$$

$$Y_3(m = 3, n = 5) = 1.4 + 1.4 = 2.8$$
$$Y_3(m = 3, n = 4) = 1.6 + 1.9 = 3.5$$
$$Y_3(m = 3, n = 3) = 1.9 + 0.7 = 2.6.$$

The cost-to-go values are shown in Fig. 4.3 in correspondence of the respective path. Thus, the best path from E to L passes through K and has a cost-to-go of 2.5 (i.e., $1.8 + 0.7$); the best path from F to L passes through K and has a cost-to-go of 1.6, and the best path from G to L passes through K and has a cost-to-go of 2.6. This is all the information needed before the algorithm moves to the preceding time step ($k = 2$), and computes the arc costs for points B, C, and D. Because of Bellman's optimality principle, the optimal path from E, F, or G to L is not affected by the choice at the previous time step, therefore the cost-to-go from B to L is the sum of the arc cost from B to either E, F, G, and of the optimal cost from there to L: for example, going from B to L passing through E costs 4.4, i.e., 1.9 (cost of B→E) plus 2.5 (lowest cost of E→L). With similar reasoning, the entire graph of Fig. 4.3 is completed with the cost-to-go values $Y(x, k)$:

$$Y(x, k) = \begin{bmatrix} 3 & 2.5 & 1.4 \\ 4.8 & 4.1 & 1.6 & 1.9 \\ 3.3 & 2.6 & 0.7 \end{bmatrix}$$

The corresponding optimal control matrix μ^*, which contains the control values to be applied at each time step (column) and for each state (row) is the following:

$$\mu^*(x, k) = \begin{bmatrix} -0.05 & -0.1 & -0.05 \\ +0.05 & -0.05 & -0.05 & 0 \\ +0.05 & 0 & +0.05 \end{bmatrix}$$

At this point, it is possible to choose the optimal path as the one that gives the lowest total cost from A to L. Its value is $J^*(x_0) = Y(x_0, 1) = 4.8$ and the optimal policy from A to L, which is obtained by reading the values of $\mu^*(x, k)$ according to the last box of Fig. 4.1, is:

$$u^* = \{+0.05, \ -0.05, \ -0.05, \ +0.05\}$$

(which corresponds to the bold path in Fig. 4.3).

DP provides a numerical optimal solution, within the accuracy limits due to the discretization of the candidate solutions. However, it is not applicable in real time, for two important reasons:

1. the solution has to be calculated backward, therefore the entire driving cycle must be known a priori, and
2. it is a procedure computationally heavy, requiring the backward solution of the entire problem before being able to determine the first control action.

However, dynamic programming provides the closest approximation to the optimal solution of the energy management problem, and is often used to determine the maximum potentiality of a given architecture, thus serving as a design tool or as a benchmark for implementable control strategies [4–12].

DP implementation as described is a three-step process: determination of arc costs, minimization of cost-to-go to determine the optimal control policy (proceeding backward in time), and application of the optimal control policy to the system (proceeding forward in time). While the latter step can be performed on a standard simulator, the first two steps require a specific coding able to separate clearly the definition of the cost at each time from the integration of the dynamic state equations. In other words, a non-dynamic representation of the cost function $L_k = L(x_k, u_k, w_k)$

and the system function $x_{k+1} = f_k(x_k, u_k, w_k)$ are needed, where the external disturbance w_k may represent the driving cycle, the road slope, etc. These functions are then called by the DP algorithm to compute the cost-to-go starting from the last point of the driving cycle, and to determine the optimal policy.

The reader is directed to [13, 14] for a detailed description of a general-purpose DP algorithm that is made available for download [15].

References

1. D. Bertsekas, *Dynamic Programming and Optimal Control* (Athena Scientific, Belmont, 1995)
2. R.E. Bellman, *Dynamic Programming* (Princeton University Press, Princeton, 1957)
3. R. Weber, *Optimization and Control*. Course notes (University of Cambridge, Cambridge, 2007)
4. A. Brahma, Y. Guezennec, G. Rizzoni, Optimal energy management in series hybrid electric vehicles, in *Proceedings of the 2000 American Control Conference*, vol. 1, no. 6 (2000), pp. 60–64
5. C. Lin, J. Kang, J. Grizzle, H. Peng, Energy management strategy for a parallel hybrid electric truck, in *Proceedings of the 2001 American Control Conference*, vol. 4 (2001), pp. 2878–2883
6. C. Lin, Z. Filipi, Y. Wang, L. Louca, H. Peng, D. Assanis, J. Stein, Integrated, feed-forward hybrid electric vehicle Simulation in simulink and its use for power management studies. SAE paper 2001–01–1334 (2001)
7. C. Lin, H. Peng, J. Grizzle, J. Kang, Power management strategy for a parallel hybrid electric truck. IEEE Trans. Control Syst. Technol. **11**(6), 839–849 (2003)
8. A. Sciarretta, M. Back, L. Guzzella, Optimal control of parallel hybrid electric vehicles. IEEE Trans. Control Syst. Technol. **12**(3), 352–363 (2004)
9. L. Pérez, G. Bossio, D. Moitre, G. García, Optimization of power management in an hybrid electric vehicle using dynamic programming. Math. Comput. Simul. **73**(1–4), 244–254 (2006)
10. T. Hofman, M. Steinbuch, R. van Druten, A. Serrarens, Rule-based energy management strategies for hybrid vehicles. Int. J. Electr. Hybrid Veh. **1**(1), 71–94 (2007)
11. D. Bianchi, L. Rolando, L. Serrao, S. Onori, G. Rizzoni, N. Al-Khayat, T. Hsieh, P. Kang, A rule-based strategy for a series/parallel hybrid electric vehicle: an approach based on dynamic programming, in *Proceedings of 2010 ASME Dynamic Systems and Control Conference* (2010)
12. L. Serrao, S. Onori, G. Rizzoni, A comparative analysis of energy management strategies for hybrid electric vehicles. ASME J. Dyn. Syst. Meas. Control **133**(3), 031012 (2011)
13. O. Sundström, L. Guzzella, A generic dynamic programming Matlab function, in *18th IEEE International Conference on Control Applications (CCA) & Intelligent Control (ISIC)* (IEEE, 2009), pp. 1625–1630
14. O. Sundström, D. Ambühl, L. Guzzella, On implementation of dynamic programming for optimal control problems with final state constraints. Oil Gas Sci. Technol.-Rev., IFP (2009)
15. ETH Zurich, Institute for Dynamic Systems and Control. DPM function. http://www.idsc.ethz. ch/Downloads/DownloadFiles/dpm

Chapter 5
Pontryagin's Minimum Principle

5.1 Introduction

In Chap. 4 we presented the DP as a numerical tool to solve the optimal control problem for hybrid electric vehicles as defined in Sect. 3.4. On the one hand, this technique provides the optimal global solution to the problem; on the other hand, the curse of dimensionality from which this algorithm suffers along with the need of a backward discretized model do not make it appealing from a computational and implementation standpoint [1, 2]. For, other optimal control methods have been applied to solve this type of problem, among which Pontryagin's minimum principle[1] [1, 3, 5] is probably the one that has received the most attention in recent years.

Pontryagin's minimum principle is in the form of a set of necessary conditions of optimality. A control law $\mathbf{u}(t)$ that satisfies the conditions of the minimum principle is called *extremal*. Being the conditions of the minimum principle only necessary, the optimal solution, when one exists, must be an extremal control. Conversely, not all extremal controls are optimal.

There are several formulations of the principle, depending on the way the optimal control problem is specified [5]. The most relevant for the HEV energy management problem is discussed in this chapter.[2] In Sect. 5.2 the formulation of the principle is given for the general case of a nth dimensional multi-input system, whereas in Sect. 5.3 the principle is presented in detail for the case of HEV energy management. The chapter closes with some remarks about the relationship between dynamic programming and the minimum principle.

---

[1]The minimum principle was originally proposed (as *maximum* principle) by the Russian mathematician Lev Semenovich Pontryagin and his students in 1958 and later described in a textbook [3]. Some regard this theorem as the beginning of modern optimal control theory [4].

[2]The reader can refer to [5] for other PMP formulations.

© The Author(s) 2016
S. Onori et al., *Hybrid Electric Vehicles*, SpringerBriefs in Control,
Automation and Robotics, DOI 10.1007/978-1-4471-6781-5_5

5.2   Minimum Principle for Problems with Constraints on the State

The formulation of Pontryagin's minimum principle used when solving Problem 3.2 defined in Sect. 3.5 is the one related to the case including constraints on the system states.

As discussed in Chap. 3, the state variables are constrained to remain within some boundaries, in general time-varying: $\mathbf{x}(t) \in \Omega_{\mathbf{x}}(t) \subset \mathbb{R}^n \ \forall \, t \in [t_0, \, t_f]$. Formally, the state boundaries can be expressed by defining the set of admissible states as those for which the conditions $\mathbf{G}(\mathbf{x(t)}) \leq \mathbf{0}$ are satisfied, i.e.:

$$\Omega_{\mathbf{x}}(t) = \{\mathbf{x} \in \mathbb{R}^n | G(\mathbf{x}(t)) \leq 0\},$$

where the function $\mathbf{G}(\mathbf{x(t)}) : \mathbb{R}^n \mapsto \mathbb{R}^m$ represents a set of m inequalities that the components of the state vector must satisfy.

For the Problem 3.2, the Hamiltonian function is defined as

$$H\left(\mathbf{x}(t), \mathbf{u}(t), \boldsymbol{\lambda}(t), t\right) = L\left(\mathbf{x}(t), \mathbf{u}(t), t\right) + \boldsymbol{\lambda}(t)^T \cdot f\left(\mathbf{x}(t), \mathbf{u}(t), t\right) \tag{5.1}$$

where $f\left(\mathbf{x}(t), \mathbf{u}(t), t\right)$ is the right-hand side of the system dynamic equation (3.9), $L\left(\mathbf{x}(t), \mathbf{u}(t), t\right)$ is the instantaneous cost in (3.13), and $\boldsymbol{\lambda}(t)$ a vector of optimization variables, also known as *adjoint states* or *co-states* of the system. The co-state vector $\boldsymbol{\lambda}(t)$ has the same dimension as the state vector $\mathbf{x}(t)$.

Pontryagin's minimum principle states that if $\mathbf{u}^*(t)$ is the optimal control law for Problem 3.2 then the following conditions are satisfied [5]:

1. the state and co-state must satisfy the following conditions:

$$\dot{\mathbf{x}}^*(t) = \left. \frac{\partial H}{\partial \boldsymbol{\lambda}} \right|_{\mathbf{u}^*(t)} = f\left(\mathbf{x}^*(t), \mathbf{u}^*(t), t\right) \tag{5.2}$$

$$\dot{\boldsymbol{\lambda}}^*(t) = - \left. \frac{\partial H}{\partial \mathbf{x}} \right|_{\mathbf{u}^*(t)} = h\left(\mathbf{x}^*(t), \mathbf{u}^*(t), \boldsymbol{\lambda}^*(t), t\right)$$

$$= -\frac{\partial L}{\partial \mathbf{x}}\left(\mathbf{x}^*(t), \mathbf{u}^*(t), t\right) - \boldsymbol{\lambda}^*(t) \cdot \left[\frac{\partial f}{\partial \mathbf{x}}\left(\mathbf{x}^*(t), \mathbf{u}^*(t), t\right)\right]^T \tag{5.3}$$

$$\mathbf{x}^*(t_0) = x_0 \tag{5.4}$$

$$\mathbf{x}^*(t_f) = x_{target} \tag{5.5}$$

2. for all $t \in [t_0, \, t_f]$, $\mathbf{u}^*(t)$ globally minimizes the Hamiltonian:

$$H(\mathbf{u}(t), \mathbf{x}^*(t), \boldsymbol{\lambda}^*(t), t) \geq H(\mathbf{u}^*(t), \mathbf{x}^*(t), \boldsymbol{\lambda}^*(t), t), \quad \forall \, \mathbf{u}(t) \in U(t), \forall \, t \in [t_0, \quad , t_f]$$

i.e., the optimal solution $\mathbf{u}^*(t)$ is such that

$$\mathbf{u}^*(t) = \arg \min_{\mathbf{u}(t) \in U(t)} (H(\mathbf{u}(t), \mathbf{x}(t), \lambda(t), t)) \tag{5.6}$$

where $U(t)$ indicates the set of admissible control values at time t.

5.2.1 On the System State Boundaries

For the problem at hand, the constraints on the system state are both local and global. Global constraints are guaranteed by ensuring that both (5.4) and (5.5) are satisfied. Local constraints on the system states, to guarantee that the states of the system are in the region $\Omega_x(t)$ at each instant of time, can be enforced by introducing an extra cost in the Hamiltonian that is activated whenever the state boundaries are reached or violated.

Because of the constraints on the state variable, the formulation of the principle depends on whether the state constraints are active (i.e., the state assumes a boundary value) or not. In particular, the constraints on the states are violated when (from (3.14) and (3.15)):

1 the state assumes values above its admissible upper limits:

$$G_1(\mathbf{x}(t)) = \mathbf{x}(t) - \mathbf{x}_{max} > 0 \tag{5.7}$$

2 or, the state assumes values below its admissible lower limits:

$$G_2(\mathbf{x}(t)) = \mathbf{x}_{min} - \mathbf{x}(t) > 0 \tag{5.8}$$

In order to introduce these constraints formally in the formulation of the principle, the total time derivatives $G^{(r)}$ of G_1 and G_2 are used, up to the order r in which $\mathbf{u}(t)$ appears explicitly for the first time. For the problem dealt with in this book, the control $\mathbf{u}(t)$ appears in the first time derivative of G_1 and G_2, namely $r = 1$.

$$\mathbf{G}^{(1)}(\mathbf{x}(t), \mathbf{u}(t), t) = \begin{cases} G_1^{(1)}(\mathbf{x}(t), \mathbf{u}(t), t) = \frac{dG_1}{dt} = \dot{\mathbf{x}}(t) = f(\mathbf{x}(t), \mathbf{u}(t), t) \\ G_2^{(1)}(\mathbf{x}(t), \mathbf{u}(t), t) = \frac{dG_2}{dt} = -\dot{\mathbf{x}}(t) = -f(\mathbf{x}(t), \mathbf{u}(t), t) \end{cases}$$
$$\tag{5.9}$$

Using (5.9), the new Hamiltonian function enforcing the local state constraints is defined as [5]:

$$H = L(\mathbf{x}(t), \mathbf{u}(t), t) + \lambda(t)^T \cdot f(\mathbf{x}(t), \mathbf{u}(t), t) + w(\mathbf{x})^T \cdot f(\mathbf{x}(t), \mathbf{u}(t), t) \tag{5.10}$$

where

$$w(\mathbf{x}) = \begin{cases} 0 & \text{if } \mathbf{G}\left(\mathbf{x}(t)\right) < 0 \text{ (constraints not active)} \\ -K & \text{if } G_1\left(\mathbf{x}(t)\right) \geq 0 \text{ (upper constraints active)} \\ K & \text{if } G_2\left(\mathbf{x}(t)\right) \geq 0 \text{ (lower constraints active)} \end{cases} \qquad (5.11)$$

$w(\mathbf{x})$ has the same number of components as $\mathbf{G}(\mathbf{x}(t))$ and each component is defined on the basis of the corresponding component of $\mathbf{G}(\mathbf{x}(t))$. The value of the constant K is arbitrary. A general rule is to make K high enough as to guarantee that the additional cost due to meeting (or exceeding) a state constraint makes the corresponding solution unacceptable.

The Hamiltonian function (5.10) can be rewritten as:

$$H = L\left(\mathbf{x}(t), \mathbf{u}(t), t\right) + \left(\boldsymbol{\lambda}(t)^T + w(\mathbf{x})^T\right) \cdot f\left(\mathbf{x}(t), \mathbf{u}(t), t\right) \qquad (5.12)$$

Practically, the Hamiltonian function during the intervals in which the constraints are active is augmented by the term $w(\mathbf{x})$, called *additive penalty function* that depends on the derivative of the constraint function. For constraints of the form (5.7) or (5.8), the effect of this is that control candidates that tend to overcome the state boundaries become penalized, while those of opposite sign are favored.

In light of the newly defined Hamiltonian (5.12), the co-state equation (5.3) is modified as follows:

$$\dot{\boldsymbol{\lambda}}^*(t) = -\frac{\partial L}{\partial \mathbf{x}}\left(\mathbf{x}^*(t), \mathbf{u}^*(t), t\right) - \left(\boldsymbol{\lambda}^*(t) + w(\mathbf{x})\right)\left[\frac{\partial f}{\partial \mathbf{x}}\left(\mathbf{x}^*(t), \mathbf{u}^*(t), t\right)\right]^T. \quad (5.13)$$

The inclusion of the additive penalty function introduces discontinuities in the Hamiltonian function at the time instants in which the state boundaries are reached. This translates into discontinuities in the value of the co-state $\lambda(t)$ at those instants, as can be inferred from (5.13).

5.2.2 Notes on the Minimum Principle

Pontryagin's minimum principle is a rather powerful tool to solve finite horizon optimization problems. It permits redefining the global optimal control problem in terms of local conditions expressed by the differential equations (5.2) and (5.13) and by the instantaneous minimization (5.6). Clearly, the global nature of the problem does not disappear, and remains evident in the fact that the boundary conditions are given at the initial and final time; therefore, the problem cannot be solved as a standard dynamic evolution problem.

The necessary conditions given by the Pontryagin's minimum principle can be used to find optimal control candidates, called *extremal controls*; Pontryagin's principle ensures that the optimal control, if it exists, must be an extremal control. If the optimal control problem admits a solution, and there is only one extremal control, then that is the optimal control solution. Even if several extremal controls are found, it may be relatively easy to simply apply all of them one at a time and then identify the optimal control as the *extremal* giving the lowest total cost.

A standard way of solving an optimal control problem using the minimum principle is the so called *shooting method*[3] which consists in applying an arbitrary initial value (guess) of the co-state at the beginning of the simulation, λ_0, then running the dynamic problem defined by (5.2) and (5.3) while solving the minimization (5.6) at each time instant. At the end of the optimization horizon, the state and co-state will reach a final value which may not satisfy the terminal constraints. In this case, the initial value of λ is changed and the entire problem is run again until a suitable initial value λ_0^* is found, which generates a solution meeting all the constraints.

5.3 Pontryagin's Minimum Principle for the Energy Management Problem in HEVs

The HEV optimal energy management problem is a scalar problem both in the state and in the control, i.e., $n = 1, p = 1$. The state, SOC, must be between two values, SOC_{max} and SOC_{min}. Thus the set of admissible states is: $\Omega_{SOC}(t) = [SOC_{min}, SOC_{max}]$. The control, $P_{batt}(t)$, lives in the set of admissible control values $U_{P_{batt}}(t) = [P_{batt,min}(t), \ P_{batt,max}(t)]$.

The explicit dependence of the Hamiltonian on time t, in (5.1) or (5.10) translates into a dependence on the power request from the driver, P_{req}. Hence, the Hamiltonian function for the HEV energy management problem is:

$$H\left(SOC(t), P_{batt}(t), \lambda(t), P_{req}(t)\right) \tag{5.14}$$
$$= \dot{m}_f\left(P_{batt}(t), P_{req}(t)\right) + (\lambda(t) + w(SOC)) \cdot \dot{SOC}(t),$$

and the necessary conditions are:

$$P_{batt}^*(t) = \arg\min_{P_{batt}(t) \in U_{P_{batt}}} H\left(P_{batt}(t), SOC(t), \lambda(t), P_{req}(t)\right) \tag{5.15}$$

$$\dot{SOC}^*(t) = f(SOC^*(t), P_{batt}^*(t)) \tag{5.16}$$

$$\dot{\lambda}^*(t) = -\left(\lambda^*(t) + w(SOC)\right)\frac{\partial f}{\partial SOC}(SOC^*, P_{batt}^*) = h(SOC^*(t), P_{batt}^*(t), \lambda^*(t)) \tag{5.17}$$

$$SOC^*(t_0) = SOC_0 \tag{5.18}$$

$$SOC^*(t_f) = SOC_{target} \tag{5.19}$$

---

[3]This approach is practical and reliable when the problem has a single state and the effect of the co-state on the solution is easily understood; in that case, the shooting method can be implemented with a simple iterative search, such as bisection, which converges in a relatively few steps.

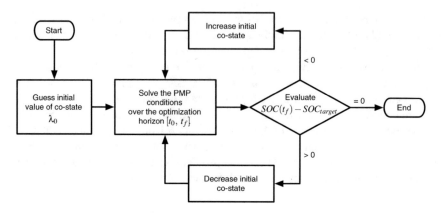

Fig. 5.1 Iterative mechanism to solve the Pontryagin's minimum principle via shooting method

$$SOC_{min} \leq SOC^*(t) \leq SOC_{max} \tag{5.20}$$

Equations (5.16) and (5.17) represent a system of two first-order differential equations in the variables SOC and λ. Despite being completely defined, this two-point boundary value problem can be solved numerically only using an iterative procedure, because one of the boundary conditions is defined at the final time, $SOC^*(t_f) = SOC_{target}$.

The solution of the PMP's necessary conditions is obtained via shooting method according to the scheme in Fig. 5.1.

Starting with an initial guess of λ_0, at each iteration of the shooting method the minimum principle conditions are solved throughout the length of the optimization horizon, $[t_0, \ t_f]$, typically corresponding to the duration of a driving cycle. At the end of the simulation, the obtained value of the $SOC(t_f)$ is compared to the desired state of charge, SOC_{target}. Depending on the difference $SOC(t_f) - SOC_{target}$, the value of λ_0 is either adjusted and the simulation repeated, or the algorithm ends if the difference reaches the desired target (i.e., it is close to zero within a pre-defined tolerance). A bisection procedure can be used to obtain convergence in few iterations, making the minimum principle sensibly faster than dynamic programming.

The implementation of the PMP's necessary conditions is shown in the schematic of Fig. 5.2. At each instant of time over the optimization horizon $[t_0, \ t_f]$, given a request of power, P_{req}, the Hamiltonian is built and minimized. This generates the optimal control, $P_{batt}^*(t)$ that is applied to the state and co-state dynamic block to compute the state of charge and co-state variation at the next step.

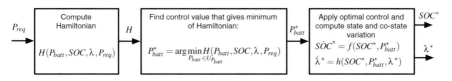

Fig. 5.2 Open-loop PMP-based energy management control scheme

It is interesting to notice that in the scalar co-state equation (5.17), the term $\frac{\partial f}{\partial SOC}$ (SOC^*, P_{batt}^*) can be expanded using (3.4), as follows:

$$\frac{\partial f(SOC, P_{batt})}{\partial SOC} = -\frac{1}{\eta_{coul}^{sign(I(t))} Q_{nom}} \frac{\partial I(V_{oc}(SOC), R_0(SOC), P_{batt})}{\partial SOC}$$

$$= -\frac{1}{\eta_{coul}^{sign(I(t))} Q_{nom}} \left[\frac{\partial I}{\partial V_{oc}} \frac{\partial V_{oc}}{\partial SOC} + \frac{\partial I}{\partial R_0} \frac{\partial R_0}{\partial SOC} \right] \qquad (5.21)$$

It is often the case that the $V_{oc}(SOC)$ and $R_0(SOC)$ characteristics of the battery are such that their dependence on SOC (over the SOC range of operation) can be neglected. This leads to

$$\frac{\partial f}{\partial SOC}(SOC^*, P_{batt}^*) \approx 0$$

Thus, when the optimal control operates within the state boundaries the co-state equation (5.17) can be approximated to $\dot{\lambda} \approx 0$.

The implication of this fact is in that the PMP optimal solution of the HEV energy management problem is characterized by a constant co-state λ. This constant, though, is unknown when attempting to solve the problem and it can be found by means of the *shooting method*, as mentioned earlier in this chapter.

The additive penalty function $w(SOC)$ is the piecewise function shown in Fig. 5.3 and given by:

$$w(SOC) = \begin{cases} 0 & \text{if } SOC_{max} < SOC < SOC_{min} \\ K & \text{if } SOC < SOC_{min} \\ -K & \text{if } SOC > SOC_{max} \end{cases} \qquad (5.22)$$

The constant K is determined in simulation iteratively by trial-and-error to ensure that the cost of using the battery becomes high whenever the SOC hits the lower bound SOC_{min} and low whenever the SOC hits the upper bound SOC_{max}. When SOC is within its maximum and minimum allowable bounds, the penalty function is not active adding a zero term to the instantaneous cost.

Fig. 5.3 Piecewise-constant
penalty function $w(SOC)$

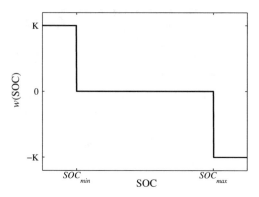

5.3.1 Power-Based PMP Formulation

For a more immediate interpretation of the Hamiltonian and without any loss of generality,[4] the PMP can be reformulated by using the electrochemical energy variation $E_{ech}(t)$ as the system state instead of SOC, and the power equivalent to the fuel flow rate, P_{fuel}, as the instantaneous cost L in (3.13), instead of the fuel mass flow rate $\dot{m}_f(t)$.

Let us first define the battery electrochemical energy variation as the amount of energy that is extracted from the battery during a given time interval $[t_0,\ t]$. This is:

$$E_{ech}(t) = \int_{t_0}^{t} P_{ech}(t)dt = \int_{t_0}^{t} V_{oc}(SOC)I(t)dt. \qquad (5.23)$$

where $P_{ech}(t)$ is the *battery electrochemical power*, representing the power that corresponds to the effective battery discharge/charge. In a charge-sustaining HEV, the limited SOC sweep results in working at approximately constant potential, i.e., $V_{oc} \approx const = V_{oc,nom}$ (nominal value of the open circuit voltage). Under this assumption and using (2.33) and (5.23) can be expressed as follows:

$$E_{ech}(t) = V_{oc,nom} \int_{t_0}^{t} I(t)dt \qquad (5.24)$$

$$= V_{oc,nom} \cdot Q_{nom} \cdot \eta_{coul} \cdot (SOC(t_0) - SOC(t)). \qquad (5.25)$$

The power equivalent to the fuel flow rate, on the other hand, is:

$$P_{fuel}(t) = Q_{lhv}\dot{m}_f(t). \qquad (5.26)$$

---

[4]As well as for further developments in the next chapters.

Taking the derivative with respect to time of (5.25) produces the state equation:

$$P_{ech}(t) = \dot{E}_{ech}(t) = -\dot{SOC}(t)V_{oc,nom}Q_{nom}\eta_{coul} = V_{oc,nom}I(t) \qquad (5.27)$$

The Hamiltonian becomes

$$H = P_{fuel}(t) + \lambda(t) \cdot P_{ech}(t) \qquad (5.28)$$

or

$$H = P_{fuel}(t) + \big(\lambda(t) + w(SOC)\big) \cdot P_{ech}(t) \qquad (5.29)$$

to include state boundary constraints. Equation (5.28) can be interpreted as an equivalent power where $\lambda(t)$ represents a weighting factor that transforms the battery power into fuel power. With this PMP formulation λ is an adimensional factor, as opposed to the formulation given by (5.14) where λ carries the units of grams.

The co-state $\lambda(t)$, using the new state variable $E_{ech}(t)$, evolves according to:

$$\dot{\lambda}(t) = -\frac{\partial H}{\partial E_{ech}} = -\big(\lambda(t) + w(SOC)\big)\frac{\partial P_{ech}}{\partial E_{ech}} \qquad (5.30)$$

which, by virtue of (5.25) and (5.27), yields to the same co-state equation (5.17) used in the original formulation.

The existence and uniqueness of the solution cannot be proved formally in the general case, but it is reasonable to assume that at least one optimal solution exists for the energy management problem, in the sense that there must necessarily be at least one sequence of controls giving the lowest possible fuel consumption. If the minimum principle generates only one extremal solution, that can be considered the optimal solution; if there is more than one extremal solution, they are all compared (i.e., the total cost resulting from the application of each is evaluated) and the one yielding the lowest total cost is chosen. In [6] it is shown that under the assumption of constant battery efficiency the PMP necessary conditions are also sufficient.

Simulations results showing the PMP implementation are discussed at length in Chap. 8.

A well-known fact about the PMP solution is the high sensitivity of optimality to the initial co-state value [5, 7]. For, Figs. 5.4 and 5.5 show the SOC variation obtained by using different initial values of the co-state. Increasing the absolute value of λ results in a tendency to increase the SOC during the driving cycle, and vice-versa. For each driving cycle, there exists one value of co-state for which the solution is charge-sustaining, which is easily found iteratively thanks to the predictable behavior visible in Fig. 5.5. A numerical example detailing the search procedure is discussed in Sect. 8.2.4.

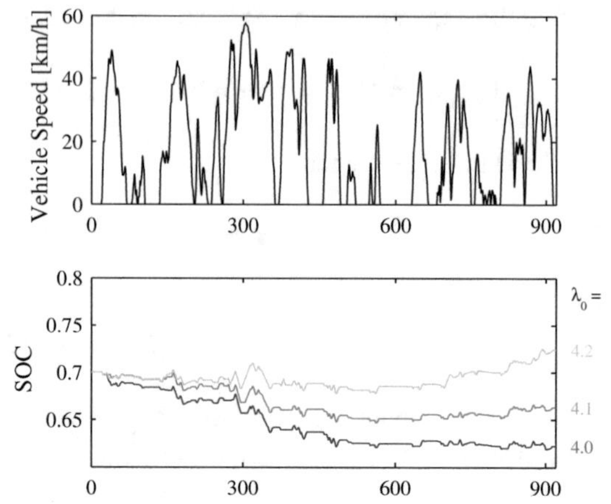

Fig. 5.4 Effect of initial co-state value on *SOC* evolution (simulation results obtained from the case study described in Sect. 8.2, cycle Artemis Urban)

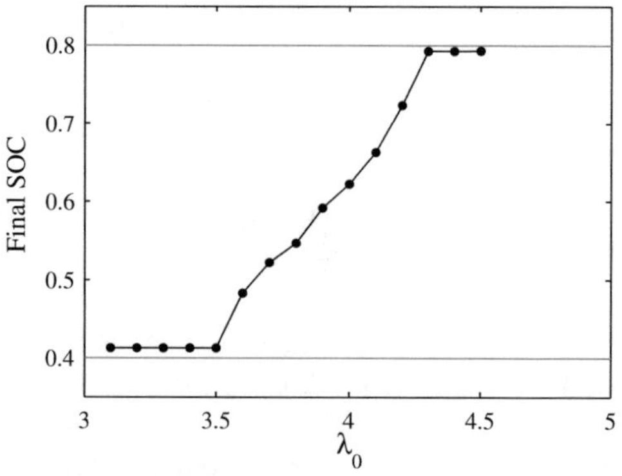

Fig. 5.5 Effect of initial co-state value on final *SOC* value (simulation results obtained from the case study described in Sect. 8.2, cycle Artemis Urban)

5.4   Co-State λ and Cost-to-Go Function

The aim of this section is to qualitatively discuss the relationship between the DP method, presented in Chap. 4, and the PMP solution discussed in this chapter. In order to do that, we make use of the Hamilton–Jacobi–Bellman (HJB) equation which is a partial differential equation central to optimal control theory. The interested reader is referred to [1, 8, 9] for the mathematical formulation of HJB.

The solution of the HJB equation is the "value function," which gives the optimal cost-to-go for a given dynamical system with an associated cost function. The corresponding discrete-time equation is the Bellman equation presented in Chap. 4.

From the theory of calculus of variations [8, 9], it is known that if the optimal control problem is solved and the optimal value of the cost objective is known, J^*, then this defines the optimal trajectory from the initial state x_0 to the final state x_f. For simplicity, we limit our discussion to the case $n = 1$ and $p = 1$. Based on Bellman's

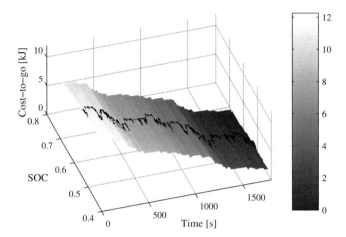

Fig. 5.6 Example of cost-to-go $J^*(x, t)$ (in kJ) computed for the case study presented in Sect. 8.2, and the optimal *SOC* trajectory superposed

principle of optimality [1], the HJB equation can be derived that states that, on the optimal trajectory, the following holds true:

$$-\frac{\partial J^*}{\partial t} = \min_{u(t)} \left[\dot{L}(x, u) + \frac{\partial J^*}{\partial x} f(x, u) \right] \tag{5.31}$$

Given (5.31), a connection can be made between the HJB and PMP along the optimal trajectory. Providing the minimization problem is solved, i.e., the optimal cost function $J^*(x)$ is found, then the optimal trajectory is obtained [10].

Evaluating the partial derivative of $J^*(x)$ with respect to x, along the optimal trajectory x^*, gives:

$$\lambda^*(t) = \frac{\partial J^*}{\partial x} \bigg|_{x^*} \tag{5.32}$$

Thus, supposing that the minimization problem is solved (e.g. by means of DP) and that $J^*(x)$ is known, this defines the optimal trajectory in terms of state $x^*(t)$ and also in terms of co-state $\lambda^*(t)$.

As an example of application of such considerations to the HEV case (as first proposed in [10]), Fig. 5.6 shows results obtained from the DP solution of the problem. The cost-to-go map obtained from DP is shown, as well as the optimal *SOC* trajectory. The co-state computed numerically using (5.32) is shown in Fig. 5.7 for the entire domain. Figure 5.8 shows sections of the surface for several time values, in order to show the clear trend of the co-state variation with respect to the system state (*SOC* is shown here).

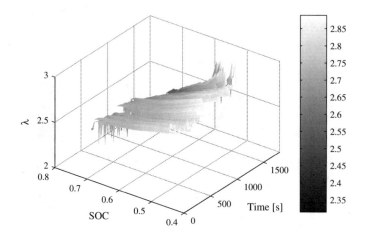

Fig. 5.7 Co-state $\lambda(x, t)$ computed for the problem defined in Sect. 8.2, obtained by applying (5.32) to the cost-to-go $J^*(x, t)$ shown in Fig. 5.6

Fig. 5.8 Same surface as in Fig. 5.7, shown here in the plane (SOC, λ) for several values of time, randomly chosen, in order to show the correlation between λ and SOC

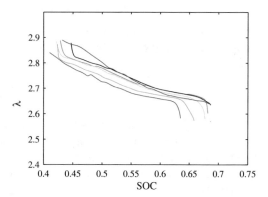

The optimal trajectory state and co-state is shown in Fig. 5.9, which also includes the values computed using PMP (with the approximation of constant co-state, whose value is chosen with the shooting method). It is apparent how the co-state value computed from DP is indeed approximately constant and matches well the constant value obtained from PMP; the optimal SOC trajectories computed with the two methods are also comparable. The small differences in the behavior of SOC and λ in the two cases are due to numerical discretization effects.

This illustrates the equivalence of DP and PMP and the use of PMP to generate an optimal solution even off-line, by searching for the optimal co-state.

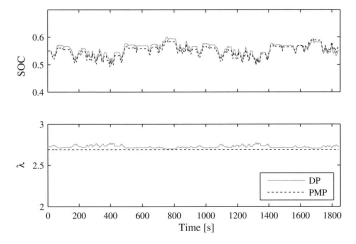

Fig. 5.9 Comparison between DP and PMP solution

References

1. D. Bertsekas, *Dynamic Programming and Optimal Control* (Athena Scientific, Belmont, 1995)
2. A. Bryson, *Dynamic Optimization* (Addison Wesley Longman, Menlo Park, 1999)
3. L. Pontryagin, V. Boltyanskii, R. Gamkrelidze, E. Mishchenko, *The Mathematical Theory of Optimal Processes* (Inderscience Publishers, New York, 1962)
4. H. Sussmann, J. Willems, 300 years of optimal control: from the brachystochrone to the maximum principle. IEEE Control Syst. Mag. **17**(3), 32–44 (1997)
5. H.P. Geering, *Optimal Control with Engineering Applications* (Springer, Heidelberg, 2007)
6. N. Kim, S.W. Cha, H. Peng, Optimal equivalent fuel consumption for hybrid electric vehicles. IEEE Trans. Control Syst. Technol. **20**(3), 817–825 (2012)
7. A. Bryson J. Optimal control: 1950–1985. IEEE Control Syst. Mag. **16**(3), 26–33 (1996)
8. D. Kirk, *Optimal Control Theory. An Introduction* (Prentice-Hall, Upper Saddle River, 1970)
9. M. Athans, P. Falb, *Optimal Control.* (McGraw-Hill, New York, 1966)
10. D. Ambühl, A. Sciarretta, C. Onder, L. Guzzella, S. Sterzing, K. Mann, D. Kraft, M. Küsell, A causal operation strategy for hybrid electric vehicles based on optimal control theory, in *Proceedings of the 4th Symposium on Hybrid Vehicles and Energy Management* (2007)

Chapter 6
Equivalent Consumption Minimization Strategy

6.1 Introduction

The Equivalent Consumption Minimization Strategy (ECMS) is a heuristic method to address the optimal control problem defined in earlier chapters, and has been shown to provide an effective solution to the HEV energy management problem. Although chronologically the ECMS solution pre-dates the material presented in earlier chapters, we choose to introduce it at this stage because of the strong connection between ECMS and formal optimal control solutions, which will become obvious very quickly. ECMS was initially introduced by Paganelli [1] in 1999 as a method to reduce the global minimization problem defined in Sect. 3.4.1 to an instantaneous minimization problem to be solved at each instant only using arguments based on actual energy flow in the powertrain.

Recent developments in the field of energy management in HEVs have shown that Pontryagin's minimum principle conditions presented in Chap. 5 are intrinsically equivalent to the ECMS method. The analysis of this equivalence is discussed in the last section of this chapter.

6.2 ECMS-Based Supervisory Control

The ECMS is based on the notion that, in charge-sustaining hybrid electric vehicles, the difference between the initial and final state of charge of the battery is very small, negligible with respect to the total energy used. This means that the electrical energy storage system is used only as an energy buffer: ultimately all energy comes from fuel, and the battery can be seen as an auxiliary, reversible fuel tank. Any stored electrical energy used during a battery discharge phase must be replenished at a later stage using fuel from the engine, or through regenerative braking.

Two cases are possible at a given operating point:

1. the battery power is positive (discharge case) at the present time; this implies that at some future time the battery will need to be recharged, resulting in some additional fuel consumption in the future. How much fuel will be required to replenish the battery to its desired energy state depends on two factors: (1) the

S. Onori et al., *Hybrid Electric Vehicles*, SpringerBriefs in Control, Automation and Robotics, DOI 10.1007/978-1-4471-6781-5_6

operating condition of the engine at the time the battery is recharged; and (2) the amount of energy that can be recovered by regenerative braking. Both factors are in turn dependent on the vehicle load, and therefore on the driving cycle.

2. the battery power is negative (charge case): the stored electrical energy will be used to alleviate the engine load required to meet the vehicle road load, implying an instantaneous fuel saving. Again, the use of electrical energy as a substitute for fuel energy depends on the load imposed by the driving cycle.

The principle underlying the ECMS approach is that a cost is assigned to the electrical energy, so that the use of electrical stored energy is made equivalent to using (or saving) a certain quantity of fuel. This cost is obviously unknown, as it depends on future vehicle behavior, but it has been shown that the cost can be related to driving conditions in a broad sense (for example, urban versus highway driving).

The concept implemented by the ECMS is illustrated in Fig. 6.1, which refers to a parallel HEV (but the concept can be applied to a series HEV—the only difference is the location of the power summation node).

In the discharge case (Fig. 6.1a), the electric motor provides mechanical power. The dotted route is related to the future return of the used electrical energy. Of course, the operating point of this recharge cannot be known a priori, and thus an approximate mean efficiency should be set. In the charge case (Fig. 6.1b), the electric motor receives mechanical energy and converts it into electrical energy stored in the

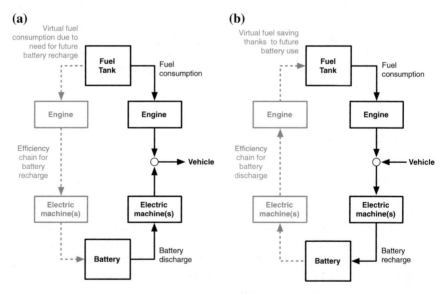

Fig. 6.1 Energy path during discharge (**a**) and charge (**b**) in a parallel HEV [2]. Practically, when implementing the ECMS we account for the use of stored electrical energy, in units of chemical fuel use (g/s), such that an equivalent fuel consumption taking into account the cost of electricity can be defined

battery. The dotted route is related to the future use of this electrical energy to produce mechanical power. This amount of mechanical energy will not have to be produced by the engine and is considered as a fuel saving. In this case, the equivalent fuel flow of the electric motor is negative.

The key idea of ECMS is that in both charge and discharge, an equivalent fuel consumption can be associated with the use of electrical energy [3]; the equivalent future (or past) fuel consumption, $\dot{m}_{ress}(t)$ (g/s), can be summed to the present real fuel consumption—fuel mass flow rate $\dot{m}_f(t)$ (g/s)—to obtain the instantaneous equivalent fuel consumption, $\dot{m}_{f,eqv}(t)$:

$$\dot{m}_{f,eqv}(t) = \dot{m}_f(t) + \dot{m}_{ress}(t). \tag{6.1}$$

By analogy to an engine which consumes *real* fuel and for which the instantaneous fuel consumption is given as

$$\dot{m}_f(t) = \frac{P_{eng}(t)}{\eta_{eng}(t)Q_{lhv}},$$

where Q_{lhv} (MJ/kg) is the fuel lower heating value (energy content per unit of mass), $\eta_{eng}(t)$ is the engine efficiency, and $P_{eng}(t)$ is the power produced by the engine when it operates at a certain efficiency, the electric machine consumes *virtual* fuel

$$\dot{m}_{ress}(t) = sfc_{eq}(t) \cdot P_{batt}(t).$$

The virtual fuel consumption can be evaluated by use of a virtual specific fuel consumption, $sfc_{eq}(t)$ (g/kWh). The virtual specific fuel consumption is proportional to an *equivalence factor* $s(t)$ which differs whether the battery is being charged or discharged:

$$\dot{m}_{ress}(t) = \frac{s(t)}{Q_{lhv}}P_{batt}(t).$$

The equivalence factor $s(t)$ is a vector of values, one for charge and one for discharge, $s(t) = [s_{chg}(t), \ s_{dis}(t)]$. Its task is to assign a cost to the use of electricity, converting electrical power into equivalent fuel consumption.

Practically speaking, the equivalence factor $s(t)$ represents the chain of efficiencies through which fuel is transformed into electrical power and vice-versa. As such, it changes for each operating condition of the powertrain. In the original formulation of ECMS, the equivalence factor is a constant, or rather a set of constants which can be interpreted as the average overall efficiency of the electric path (for each operating mode, e.g., charge or discharge) for a specific driving cycle.

Depending on the sign of P_{batt} (i.e., on whether the battery is charged or discharged), the virtual fuel flow rate can be either positive or negative, thus making the equivalent fuel consumption (6.1) higher or lower than the actual fuel consumption.

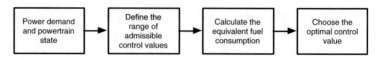

Fig. 6.2 ECMS algorithm flow

As proposed in [4], by using ECMS the global problem of minimizing the total cost is reduced to the local (instantaneous) problem of minimizing $\dot{m}_{f,eqv}(t)$:

$$Global = \begin{cases} \min_{P_{batt}(t) \in U_{P_{batt}}} \int_{t_0}^{t_f} \dot{m}_f(t)dt \\ SOC_{min} \leq SOC \leq SOC_{max} \end{cases}$$

$$\Downarrow \tag{6.2}$$

$$Local = \begin{cases} \int_{t_0}^{t_f} \min_{P_{batt}(t) \in U_{P_{batt}}} \dot{m}_{f,eqv}(t)dt \\ SOC_{min} \leq SOC \leq SOC_{max} \end{cases}$$

At each time, the equivalent fuel consumption is calculated using (6.1) for several candidate values of the control variable P_{batt}; the value that gives the lowest equivalent fuel consumption is selected.

The following steps must be executed to implement ECMS, as also illustrated in Fig. 6.2:

1. Given the state of the system in terms of $P_{req}, \omega_{eng}, \omega_{em}, SOC, \ldots$, identify the acceptable range of control $[P_{batt,min}(t), \ldots, P_{batt,max}(t)]$ which satisfies the instantaneous constraints (power, torque, current limits);
2. Discretize the interval $[P_{batt,min}(t), \ldots, P_{batt,max}(t)]$ into a finite number of control candidates;
3. Calculate the equivalent fuel consumption $\dot{m}_{f,eqv}(t)$ corresponding to each control candidate;
4. Select the control value $P_{batt}(t)$ that minimizes $\dot{m}_{f,eqv}(t)$.

Steps 1 to 4 are computed at each instant of time over the entire duration of the driving cycle.

This approach has been shown to closely approximate the global optimal solution. In addition, the instantaneous minimization problem is computationally less demanding than the global problem solved with dynamic programming, and applicable to real-world situations since it does not rely (explicitly) on information about future driving conditions.

A constant value of the equivalence factor in charge, s_{chg}, and in discharge, s_{dis}, must be selected beforehand. In practice, given a value of $s(t)$, one can pre-compute the combination of electric machine and engine power (or torque) that can meet the vehicle power (or torque) demand with minimum instantaneous equivalent fuel consumption.

The values of the equivalence factors affect the vehicle fuel consumption and the trend of the battery state of charge. The selection of the most suitable values of s_{chg}

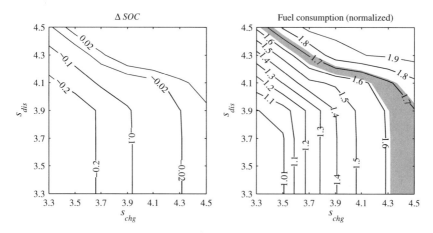

Fig. 6.3 Effect of charge and discharge equivalence factors on *SOC* variation and fuel consumption (normalized with respect to minimum value). The *shaded area* in the fuel consumption plot corresponds to charge-sustaining solutions, where $|\Delta SOC| < 0.02$ at the end of the cycle. Simulation results obtained for the case study of Sect. 8.2, cycle Artemis Urban

and s_{dis} for any given driving conditions to guarantee optimality is the challenge with ECMS.

Clearly, the concept of equivalent fuel consumption is tied with the necessity of attributing a meaningful value to the equivalence parameter. This parameter is representative of past, present, and future efficiency of the engine and the RESS, and its value affects both the charge sustainability and the effectiveness of the strategy: if it is too high, an excessive cost is attributed to the use of electrical energy and therefore the full hybridization potential is not realized; if it is too low, the opposite happens and the RESS is depleted too soon (loss of charge sustainability).

The effect of charge and discharge equivalence factors on *SOC* variation and fuel consumption is shown with a numerical example in Fig. 6.3.

Penalty Function for State Constraints and Charge Sustainability

When implementing the ECMS, a penalty function is often used to guarantee that the *SOC* does not exceed the admissible limits, $SOC_{max} \leq SOC \leq SOC_{min}$. Because of that, (6.1) is modified by using an appropriately constructed *multiplicative penalty function*, $p(SOC)$, as follows

$$\dot{m}_{f,eqv}(t) = \dot{m}_f(t) + \frac{s(t)}{Q_{lhv}} \cdot P_{batt}(t) \cdot p(SOC). \tag{6.3}$$

The multiplicative penalty function used in the instantaneous equivalent cost and shown in Fig. 6.4 is a correction function that takes into account the deviation of the current $SOC(t)$ from the target state of charge, SOC_{target}, according to the following expression

Fig. 6.4 Multiplicative
penalty function used in the
ECMS to correct for *SOC*
deviation shown for different
values of the exponent *a*. The
penalty function plays a
critical role to achieve a
reliable online estimation of
the battery state-of-charge

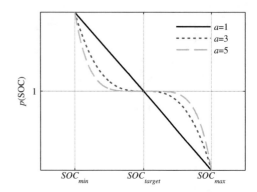

$$p(SOC) = 1 - \left(\frac{SOC(t) - SOC_{target}}{(SOC_{max} - SOC_{min})/2}\right)^{a}. \qquad (6.4)$$

The penalty function assumes unitary value if the *SOC* is at the target value
SOC_{target}, but its value changes (higher or lower) to properly compensate for the
deviation from the condition SOC_{target}. In fact, $p(SOC) < 1$ when $SOC > SOC_{target}$,
which means that a lower cost is attributed to the battery energy, thus making the
discharge more likely when the *SOC* is above the reference value. On the other hand,
$p(SOC) > 1$ when $SOC < SOC_{target}$: in this condition, the cost of battery energy is
increased to make its discharge less likely.

It has been shown [5–7] that results comparable to those achieved with dynamic
programming are obtained by using two values of the equivalence factor *s*, one for
charging (s_{chg}) and the other for discharging (s_{dis}), each of them constant during a
driving cycle. These values are different for different driving cycles and must be
obtained with a numerical optimization procedure, which is possible if the driving
cycle is known a priori. Therefore, in ideal conditions (simulation of a known cycle),
the results obtained by implementing the ECMS are very close to optimal. However,
since the strategy is very sensitive to these parameters, the control works well only
on driving cycles very similar to those in which they were obtained. This means
that, despite its "instantaneous" formulation, ECMS still implicitly relies on some
information about future driving conditions. If this information is wrong, i.e., if the
vehicle is driven on a driving cycle appreciably different than the one for which the
strategy was tuned, the control still works, but the results are not as good as they
could potentially be.

6.3 Equivalence Between Pontryagin's Minimum Principle and ECMS

The concepts behind ECMS originated from engineering intuition. However, an analytical derivation of the equivalent fuel consumption can be obtained using Pontryagin's minimum principle. In the power-based PMP formulation (Sect. 5.3.1), the Hamiltonian is

$$H = P_{fuel}(t) + \lambda(t) \cdot P_{ech}(t). \tag{6.5}$$

If we rewrite (6.1) (or (6.3)) in the ECMS in the form of power, multiplying all terms by Q_{lhv}, the instantaneous cost becomes

$$P_{eqv}(t) = P_{fuel}(t) + s(t) \cdot P_{batt}(t). \tag{6.6}$$

The similarity between (6.5) and (6.6) is clear, showing how the Hamiltonian H of the optimization problem can be regarded as an equivalent fuel consumption, or better, as an equivalent fuel power, where $\lambda(t)$ is a weighting factor that transforms the battery power into fuel power.

The battery power P_{batt} appearing in (6.6) indicates the net electrical power as seen at the battery terminals, while P_{ech} in (6.5) represents the electrochemical power, i.e., the power correlated to the effective SOC variation. If one assumes that the relation between this quantity and the electrical power can be represented as a battery charge/discharge efficiency η_{batt}, then:

$$P_{ech}(SOC(t), P_{batt}(t)) = \begin{cases} \frac{P_{batt}(t)}{\eta_{batt}(SOC, P_{batt})} & \text{if } P_{batt}(t) \geq 0 \text{ (discharge)} \\ \eta_{batt}(SOC, P_{batt})P_{batt}(t) & \text{if } P_{batt}(t) < 0 \text{ (charge)}. \end{cases} \tag{6.7}$$

Therefore, the parallel between the Hamiltonian (6.5) and the ECMS instantaneous cost (6.6) is complete if the equivalence factors in (6.6) are linked to the co-state $\lambda(t)$ as follows [8]:

$$s_{chg}(t) = \lambda(t)\eta_{batt}, \tag{6.8}$$

$$s_{dis}(t) = \frac{\lambda(t)}{\eta_{batt}}, \tag{6.9}$$

from which the following should also hold true:

$$s_{chg}(t) = \eta_{batt}^2 s_{dis}(t). \tag{6.10}$$

Thanks to the formalization based on PMP, there is no need for multiple equivalence factors, since the efficiency differences among operating conditions are implicitly taken into account in the evaluation of the quantity P_{ech}. The equivalent instantaneous fuel consumption provided by (6.5) can be used online as well as

Fig. 6.5 Effect of λ_0 on overall *SOC* variation $\Delta SOC = SOC(t_f) - SOC(t_0)$, for several driving cycles. Simulation results are based on the case study of Sect. 8.2, with the hypothesis of constant co-state

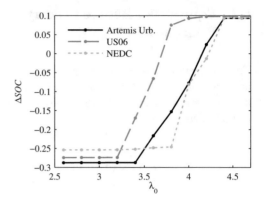

offline. In offline implementation, the optimal value of λ is found by iterative search (shooting method) as the one that satisfies all constraints. The iterative search is possible thanks to the fact that there is a direct and bi-univocal relation between the value of the co-state and the value of state of charge reached at the time t_f (as shown in Fig. 6.5).

In online implementation, where λ_0 is not known a priori, its value is adapted online to optimize the behavior based on available measurements. This is the subject discussed in the next chapter.

6.4 Correction of Fuel Consumption to Account for SOC Variation

The concept of equivalence between battery energy and fuel is also useful for the analysis of HEV fuel consumption. In fact, in practical implementations, the final *SOC* may not reach exactly the target value: therefore, in order to fairly compare fuel consumption results, it is customary to correct the actual fuel consumption value by accounting for the net amount of energy variation in the battery, i.e., always reconducing the results to a perfect charge-sustaining case. The rationale for this is easily seen in Fig. 6.6, which shows the total fuel consumption for three driving cycles, as a function of the *SOC* variation from beginning to end of the cycle.

A linear correlation between final *SOC* and fuel consumption is visible, which is easily approximated by the linear expression:

$$m_f \approx m_{f,0} + \sigma \Delta SOC \tag{6.11}$$

where m_f is the actual fuel consumption, $m_{f,0}$ is the value that would correspond to a zero *SOC* variation, and σ is a curve fitting coefficient that translates ΔSOC into a corresponding amount of fuel.

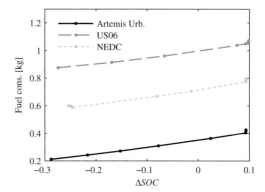

Fig. 6.6 Effect of final *SOC* on total fuel consumption, for the driving cycles of Fig. 6.5

Based on this evidence, the correction of fuel consumption can be performed by adding to the actual fuel consumption a term that depends on *SOC* variation, to obtain the charge-sustaining fuel consumption:

$$m_{f,0} = m_f - \sigma \Delta SOC. \tag{6.12}$$

The physical meaning of σ is similar to that of the ECMS equivalence factors or the PMP co-state, especially if (6.11) is written in terms of fuel energy, E_f, and battery energy, E_{ech}, rather than fuel consumption and *SOC* variation:

$$E_f = Q_{lhv} m_f \approx E_{f,0} + \bar{\sigma} \Delta E_{ech}. \tag{6.13}$$

However, σ is applied to integral measurements (energy or fuel consumption), while s or λ are weights applied to the instantaneous values of power. For this reason, their values are, generally, different although comparable: for example, Table 6.1 compares the value of σ extracted by the curves in Fig. 6.5 and the value of λ_0 found with the bisection method for the same cycles.

Table 6.1 Factor σ computed by curve fitting the linear region of the characteristics in Fig. 6.6, compared with its adimensional value $\bar{\sigma}$ obtained by expressing (6.11) in terms of energy, and to the optimal value of λ computed with the bisection method for the same cycles

Cycle	$m_{f,0}$ (kg)	σ (kg)	$\bar{\sigma} = \sigma \frac{Q_{lhv}}{E_{batt}}$ [-]	λ_0 [-]
Artemis Urb.	0.349	0.486	3.65	4.15
US 06	1.069	0.483	3.63	3.65
NEDC	0.679	0.472	3.55	4.21

The charge-sustaining fuel consumption, $m_{f,0}$, is also reported

6.5   Historical Note: One of the First Examples of ECMS Implementation

To provide an illustration of the practical application of ECMS, we briefly describe its implementation as part of one of the Advance Vehicle Technology Competitions, FutureTruck 2000, co-sponsored by the U.S. Department of Energy and General Motors. As part of the competition, a team of students and faculty demonstrated the hybridization of a model-year 2000 Chevrolet Suburban SUV, shown in Fig. 6.7 at the GM Desert Proving Grounds in Arizona, during the June 2000 competition. The production powertrain of the truck was replaced by a double-shaft parallel hybrid powertrain, shown schematically in Fig. 6.8, consisting of a 2.5 l CIDI engine (Fiat) and electric motor (Siemens) giving a combined power of 155 kW (210 hp), and coupled to the driveline by a 5-speed GM automatic transmission. Using a supervisory controller implementing ECMS, the converted SUV achieved remarkable improvements in fuel economy, while retaining most of the performance characteristics of the original vehicle. The vehicle control system architecture was based on a supervisory controller (ETAS ES-1000), which communicated the setpoints selected by the ECMS strategy to the engine control unit (Bosch ECU), the electric drive (Ford-Ecostar Traction Inverter Module, TIM), and the transmission controller (GM PCM). In addition, a custom battery interface circuit was developed to communicate between the battery management system and the supervisory controller. The energy storage system consisted of lead acid cells (Hawker Genesis) connected into a 324 V, 8 kWh battery pack.

Fig. 6.7 OSU FutureTruck 2000 at General Motors Desert Proving Ground, June 2000

Fig. 6.8 Schematic of OSU FutureTruck 2000 parallel hybrid powertrain [4]

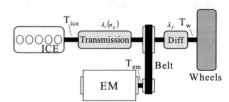

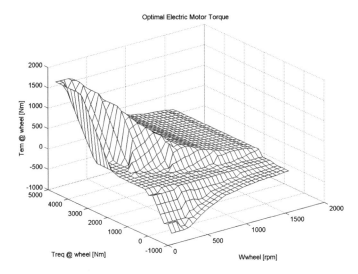

Fig. 6.9 Optimal electric machine torque map, referenced to wheels [4]

To implement ECMS in a computationally achievable form [4, 9], the equivalence factor s was assumed to be a constant and the equivalent fuel cost of the electric machine was computed for the entire range of possible electric motor operating points, on a grid. The IC engine chemical fuel cost was known from the engine brake-specific fuel consumption map, and therefore it is possible to calculate the combination of EM and ICE torques (referenced at the wheels) that would result in the minimum instantaneous fuel consumption for that particular value of s. This is done offline by computing all possible feasible solutions over a discrete grid, and selecting the combination of torques that yields the lowest value of total fuel cost. The map of Fig. 6.9 shows the optimal (minimum total fuel cost) torque contribution at the wheels for the EM (a similar map is used for the engine). The process can be repeated for different values of s, which could be representative, for example, of city and highway driving. To ensure *SOC* sustenance, and to keep the *SOC* within its target range, a penalty function as shown in Fig. 6.4 is used. For the FutureTruck 2000, the goal was to keep *SOC* between 60 and 80 %.

Figure 6.10 depicts a 22.36-miles, 46 min trip in real traffic conditions, with a mix of highway and city driving, conducted in the Columbus, OH area in August 2000. The vehicle was in a fully automated hybrid mode, embedding the supervisory control scheme described above. The only inputs to the control system were the positions of the accelerator and brake pedals. For this test, the engine was cold at the start of the test and the battery pack was fully charged.

The vertical line in Fig. 6.10, around t = 950 s, represents the boundary between the highway and the urban portions of the drive. This boundary was selected to approximate the speed statistics of the standard urban and highway US driving schedules in both portions. Actually, the highway section was driven at a speed slightly higher than the standard US FHDS. It should be noted that except for the initial battery

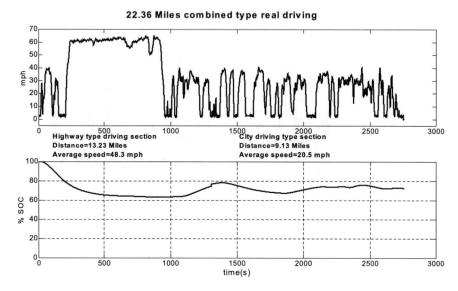

Fig. 6.10 Experimental results: driving cycle (*top*); *SOC* profile (*bottom*) [4]

transient, the *SOC* remains within the 60–80 % target range specified in the charge-sustaining operational strategy. At the very beginning, the battery *SOC* is outside this desired band (fully charged) and the control strategy automatically favors a greater use of the electric machine until the *SOC* is within the target range.

The fuel economy figures for the trip were corrected to account for the *SOC* differences between the start and end of the trip. The fuel economy on the highway section was nearly 27 mi/gal (8.8 l/100 km), and 19.7 mi/gal (12 l/100 km) in the city section. Overall fuel economy for the entire trip was 23.4 mi/gal (10.1 l/100 km). Averaging the highway and urban sections of the cycle using the conventional weights of 45 and 55 %, respectively, the overall fuel efficiency was computed to be 23 mi/gal (10.3 l/100 km). When accounting for the density and heating value difference between Diesel fuel and gasoline, the gasoline-equivalent average fuel economy was 21.2 mi/gal (11.2 l/100 km). This is approximately 1.5 times the fuel economy of the original Suburban with its 5.3 l gasoline engine. Overall tank-to-wheels efficiency for the hybrid powertrain was just below 33 %.

References

1. G. Paganelli, Conception et commande d'une chaîne de traction pour véhicule hybride parallèle thermique et électrique. Ph.D. dissertation, Université de Valenciennes, Valenciennes (1999)
2. G. Paganelli, T. Guerra, S. Delprat, J. Santin, M. Delhom, E. Combes, Simulation and assessment of power control strategies for a parallel hybrid car. Proc. Inst. Mech. Eng. Part D: J. Automob. Eng. **214**(7), 705–717 (2000)

3. G. Paganelli, G. Ercole, A. Brahma, Y. Guezennec, G. Rizzoni, A general formulation for the instantaneous control of the power split in charge-sustaining hybrid electric vehicles, in *Proceedings of 5th International Symposium on Advanced Vehicle Control*, Ann Arbor (2000)
4. G. Paganelli, G. Ercole, A. Brahma, Y. Guezennec, G. Rizzoni, General supervisory control policy for the energy optimization of charge-sustaining hybrid electric vehicles. JSAE Rev. **22**(4), 511–518 (2001)
5. A. Sciarretta, M. Back, L. Guzzella, Optimal control of parallel hybrid electric vehicles. IEEE Trans. Control Syst. Technol. **12**(3), 352–363 (2004)
6. C. Musardo, G. Rizzoni, Y. Guezennec, B. Staccia, A-ECMS: an adaptive algorithm for hybrid electric vehicle energy management. Eur. J. Control **11**(4–5), 509–524 (2005)
7. C. Musardo, B. Staccia, Energy management strategies for hybrid electric vehicles. Master's thesis, Politecnico di Milano (2003)
8. L. Serrao, S. Onori, G. Rizzoni, ECMS as a realization of Pontryagin's minimum principle for HEV control, in *Proceedings of the 2009 Conference on American Control Conference* (2009), pp. 3964–3969
9. G. Paganelli, M. Tateno, A. Brahma, G. Rizzoni, Y. Guezennec, Control development for a hybrid-electric sport-utility vehicle: strategy, implementation and field test results, in *Proceedings of 2001 American Control Conference* (2001)

Chapter 7
Adaptive Optimal Supervisory Control Methods

7.1 Introduction

The problem of designing a real-time implementable strategy to solve the energy management problem in hybrid electric vehicles and achieve a close-to-optimal solution has been the subject of extensive research over the last decade. The optimal control methods based on instantaneous minimization such as Pontryagin's Minimum Principle and Equivalent Consumption Minimization Strategy, reviewed in Chaps. 5 and 6 respectively, guarantee optimality as long as the driving cycle is perfectly known. The main challenge associated with PMP (or ECMS) is the selection of the most suitable values of λ_0 (or $[s_{ch} \ s_{dis}]$) to guarantee optimality and charge-sustainability for any given driving conditions.[1]

If perfect knowledge of the driving scenarios is not possible, PMP lends itself to suboptimal online implementable solutions providing that the co-state is suitably estimated as driving conditions change. The task of updating the co-state online as driving scenarios vary is referred to as *co-state adaptation* and the general supervisory controller is referred to as *adaptive optimal supervisory controller.*

Methods falling into this category have been indicated in the literature as Adaptive-PMP (A-PMP) or Adaptive-ECMS (A-ECMS) strategies.[2]

In this chapter, after reviewing contributions in the open literature on A-ECMS methods, we focus on and analyze methods that use a state-of-charge feedback-based mechanism to perform the co-state adaptation.

---

[1] In virtue of the equivalence between the two strategies shown in Chap. 6, in the rest of the chapter, we do not distinguish between ECMS and PMP, and we conduct the study of adaptive methods with reference to the PMP—but the same considerations apply to ECMS.

[2] In some cases [1], PMP is used to indicate the offline implementation of the minimum principle with the optimal co-state, and ECMS for its online implementation based on adaptation of λ.

© The Author(s) 2016
S. Onori et al., *Hybrid Electric Vehicles*, SpringerBriefs in Control,
Automation and Robotics, DOI 10.1007/978-1-4471-6781-5_7

7.2 Review of Adaptive Supervisory Control Methods

Before the PMP-ECMS relationship was understood, the adaptive supervisory control methods proposed last decade were aimed at using the ECMS as an online optimization strategy by properly updating the values of $[s_{ch} \; s_{dis}]$ dependently on the driving mission. In particular, two main categories of adaptation techniques to design A-ECMS can be identified:

– adaptation based on driving cycle prediction;
– adaptation based on driving pattern recognition.

After the equivalence between ECMS and PMP was formalized and a new interpretation of the ECMS was given as the optimal solution computed with PMP (see Sect. 6.3, [2] and [3, 4]), it was understood that only one parameter must be adapted for online optimization, e.g., the co-state λ. Adaptive supervisory control approaches that rely on the instantaneous minimization of the Hamiltonian and have $\lambda(t)$ as the single control parameter to adapt go under the name of A-PMP methods. The mechanism used to perform the parameter adaptation is categorized as:

– adaptation based exclusively on feedback from SOC.

Note: what follows is not meant to be an exhaustive review of contributions to the topic of adaptive strategies. Only the seminal works on the subject have been reviewed. Over the past few years, the rate of publications on this topic has exploded and it would be impossible to mention them all.

7.2.1 Adaptation Based on Driving Cycle Prediction

The driving principle behind this class of methods is: when no information on future driving conditions is available, optimal fuel economy cannot be guaranteed. Thus, this family of algorithms aims at using any sort of estimation of future information to feed the ECMS control module with the more suitable values of equivalence factors. In the early methods, described in [5] (and [6–8]), a real-time energy management strategy was proposed. It was obtained by adding to the ECMS module an on-the-fly algorithm for the estimation of the equivalence factors, where an online and periodic recalculation and optimization of the equivalence factors s_{ch} and s_{dis} was performed according to estimation of the driving conditions.

The ECMS module is thus augmented with a device able to relate the control parameter, $s(t)$, to the current velocity profile. Figure 7.1a shows the A-ECMS control diagram: the identification of the driving mission given by the *Speed Predictor* is used as input to the *Adaptor* where the best value of the equivalence factor is found based on receding-horizon optimization.

To improve the execution time, a simplification was proposed for actual real-time implementation [8], consisting in the use of one equivalence factor, for both charge and discharge, thus introducing some approximation.

(a)   **(b)**

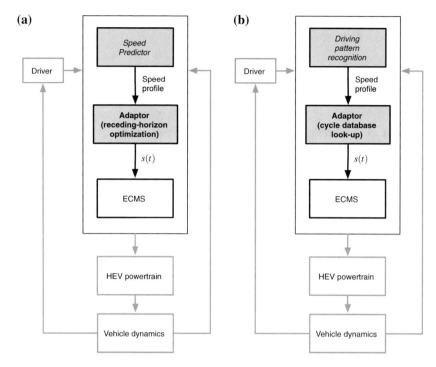

Fig. 7.1 Control diagram of A-ECMS with online optimization as proposed in [5] (**a**), with driving pattern recognition as proposed in [9] (**b**)

The performance of A-ECMS is slightly inferior to the standard ECMS tuned on a perfectly known driving cycle, but in general the results are quite good, and, most importantly, achievable in real-world application (if enough computational power is available).

In [10], the A-ECMS method strategy is based on speed prediction. The equivalence factor is estimated online based on a look-ahead horizon defined in terms of energy at the wheels, to determine at each instant the most likely behavior (charging or discharging) in the near future .

In [11] instead, an adaptation scheme similar to [6] is presented which uses a predictive reference signal generator (pRSG) in combination with a SOC tracking-based controller. The pRSG computes the desired battery SOC trajectory as a function of vehicle position such that the recuperated energy is maximized despite the constraints on the battery SOC. To compute the SOC reference trajectory, only the topographic profile of the future road segments and the corresponding average traveling speeds must be known.

In [12], the authors use a Model Predictive Control (MPC)-based strategy and utilize the information attainable from Intelligent Transportation Systems (ITS) to establish a prediction-based real-time controller structure. A constant reference SOC is considered, and A-ECMS implemented as in [6] is compared with a MPC-type controller based on the prediction of future torque demand, showing very similar performance of the two controllers.

7.2.2 Adaptation Based on Driving Pattern Recognition

In parallel to the development of adaptive supervisory control schemes based on driving cycle prediction, an alternative adaptation scheme was being proposed exploiting the fact that the equivalence factors are similar for cycles with similar statistical properties.

In [13], a multi-mode driving control algorithm using driving pattern recognition is developed and applied to a parallel HEV. The multi-mode driving control was defined as the control strategy that switches a current driving control algorithm to the algorithm optimized in a recognized driving pattern. In the study, six representative driving patterns are selected, composed of three urban driving patterns, one expressway driving pattern, and two suburban driving patterns. A total of 24 parameters such as average cycle velocity, positive acceleration, kinetic energy, stop time/total time, average acceleration, and average grade are chosen to characterize the driving patterns.

In [14] and [9], an approach for A-ECMS based on driving pattern recognition is presented to obtain better estimation of the equivalence factor in different driving conditions. A pattern recognition algorithm is used to first identify which kind of driving conditions the vehicle is undergoing, and then to select the most appropriate equivalence factors from a predefined set. The optimal values of s for several cycle typologies (city, highway, etc.) are precalculated and stored in memory (in the *equivalence factors database*); during vehicle operation, the adaptation algorithm uses the past and present driving conditions to determine the current cycle type, from which it selects the appropriate equivalency factor. The control scheme is shown in Fig. 7.1b.

While the vehicle is running, a time window of past driving conditions is analyzed periodically and recognized as one of the representative driving patterns. This operation is performed in the *Driving pattern recognition* block of Fig. 7.1b. The *Adaptor* module then selects the more suitable values of $s(t)$ from the equivalence factor database given the recognized driving patterns, and the ECMS is executed with the estimated value of $s(t)$.

7.3 Adaptation Based on Feedback from SOC

Approaches developed to design adaptive optimal supervisory control methods based on SOC feedback [15–17] are based on the idea to change dynamically the value of the co-state at the present time (without using past driving information or attempt to predict future driving behavior), in order to contrast the SOC variation and thus maintain its value around the target value. In all these methods, the SOC reference is considered constant.[3]

Performing the adaptation using a single parameter rather than two has significant advantage in that it reduces the design and calibration complexity.

Conceptually, these approaches differ in that, while [15, 16] update the equivalence factor at each time instant, [17] relies on the concept of charge-sustaining horizon, imposing charge-sustainability over a finite time horizon. If, on one hand these methods are easy to implement, robust (as they all rely on feedback from SOC) and computationally cheap, on the other hand their performance relies on a suitable tuning of the parameters used in the adaptation law.

In the next section, we analyze and compare the two main adaptive schemes based on SOC feedback from [15] and [17].

7.3.1 Analysis and Comparison of A-PMP Methods

The online adaptation of the co-state through SOC feedback uses the difference between the target state of charge, SOC_{target}, and its instantaneous value, $SOC(t)$.

In [15], an adaptation law based on a proportional-integral (PI) controller of the type:

$$\lambda(t) = \lambda_0 + k_P \left(SOC_{target} - SOC(t)\right) + k_I \int_0^t \left(SOC_{target} - SOC(\tau)\right) d\tau \quad (7.1)$$

was proposed. In (7.1), λ_0 represents the initial value of λ at time $t = 0$, and k_p and k_I are the proportional and integral gains of the adaptation law. The initialization of this algorithm, i.e., the choice of λ_0, is arbitrary, and it can be done by averaging different optimal initial values obtained offline.

In the following, (7.1) is referred to as *continuous A-PMP*. In practice, in (7.1) the integral action is added to the proportional one to guarantee better performance when tracking a constant reference value, at the price of having three tuning parameters (λ_0, k_P, k_I).

Equation (7.1) was conceived to be executed online. Such a mechanism, that is adapting the equivalence factor at each time step based on the divergence of the *SOC* from its target value, might not be always desirable, though. In fact, this continuous-time adaptation would in principle prevent using the battery over its entire range of SOC operation, as even a small deviation of the actual SOC from the constant reference value will be corrected the next time instant.

In order to allow the battery to span over a wider range of SOC, in [17], the following discrete time adaptation law (hereafter called *discrete A-PMP*) was proposed:

---

[3]Extensions have been proposed for plug-in HEVs, where the reference SOC is varied during the cycle to allow battery discharge [1, 18, 19].

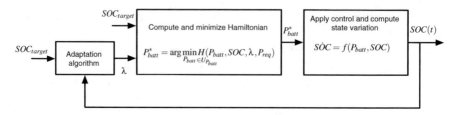

Fig. 7.2 Adaptive optimal control strategy scheme based on SOC feedback. The adaptation algorithm issues the adapted co-state λ according to either (7.1) or (7.2)

$$\lambda(k) = \frac{\lambda(k-1) + \lambda(k-2)}{2} + k_P^d\big(SOC_{target} - SOC(k)\big) \qquad (7.2)$$

where k is an integer number indicating the kth fixed time interval of length T seconds, $\lambda(k)$ is the value of the co-state in the interval $[(k-1)T, \ kT]$, and $SOC(k)$ is the value of SOC at the beginning of said interval. The concept behind (7.2) is that the charge-sustainability should be enforced over intervals of significant length (several seconds to minutes, according to battery capacity and driving cycle dynamics), thus the correction of equivalence factor should be performed only at discrete intervals and not continuously. The presence of the two previous values of λ is justified by the desire to stabilize the output. Equation (7.2) is in the form of autoregressive moving-average (ARMA) model, with two autoregressive terms and one moving average term. As mentioned, the key feature of (7.2) is that the adaptation takes place at regular intervals of duration T, rather than at each time instant, to allow for large excursion of SOC as opposed to a quasi-constant SOC trend obtained when using (7.1). While (7.2) is purely time-based, the same principle can be applied in the form of an event-based strategy, letting the adaptation take place when a certain event happens (e.g., a threshold value of SOC or speed is reached).

Regardless of how the adaption is performed (whether (7.1) or (7.2) is used), the A-PMP scheme is structured according to the feedback scheme of Fig. 7.2.

7.3.2   Calibration of Adaptive Strategies

The feedback parameters k_P and k_I of (7.1) must be tuned in order to ensure convergence of the SOC to the reference value. As for any PI controller, higher gains make the adaptation faster but potentially unstable, and the proportional and integral contribution must be suitable balanced. As an example, Fig. 7.3 shows the effect of the two gains k_P and k_I on the SOC and λ behavior, using simulation results obtained with the parallel hybrid vehicle model described in Sect. 8.2. In particular, the plots on the left-hand side show the solution as the proportional gain k_P varies while keeping the integral gain at zero, and the plots on the right-hand side show the effect of k_I when k_P is kept constant. Note how increasing k_P is not sufficient to generate a solution

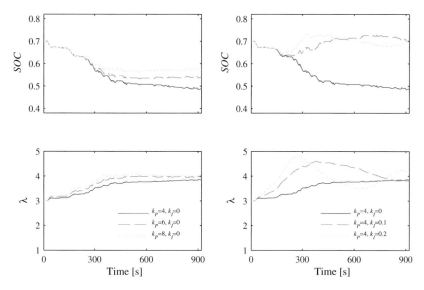

Fig. 7.3 Results obtained applying the continuous A-PMP for the cycle of Fig. 7.4 (vehicle model and characteristics as in Sect. 8.2). The co-state is adapted at each instant by using (7.1). The plots on the *left* shows the effect of the proportional gain k_P, those on the *right* the effect of the integral gain k_I

Fig. 7.4 Artemis Urban driving cycle

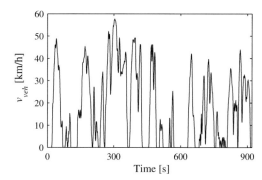

that converges to the reference value, but adding the contribution of k_I immediately helps in this respect (although an excessive integral gain generates oscillations).

Similar considerations can be observed for the discrete adaptation method (7.2). Figure 7.5 shows the effect of the gain k_P^d and of the adaptation interval T on the SOC and λ behavior.

Although the SOC behavior differs significantly, the overall fuel consumption is generally not affected as much (when accounting for the SOC variation using (6.12)), as long as the SOC boundaries are not reached: when they are, on the other hand, the battery cannot be used and this may be detrimental to the overall efficiency (for example, by preventing braking energy to be recuperated). Table 7.1 shows the SOC variation and corrected fuel consumption for each adaptive strategy and combination of calibration parameters.

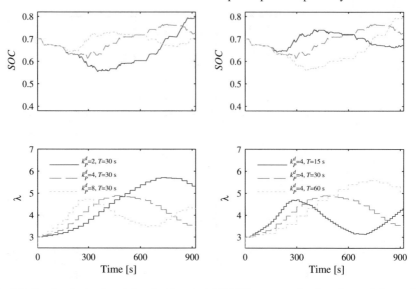

Fig. 7.5 Results obtained applying the discrete A-PMP. The co-state is adapted at each instant by using (7.2). The figure shows the effect of the gain k_P^d (plots on the *left*), and of the update interval T (plots on the *right*)

Table 7.1 Effect of calibration parameters on continuous and discrete A-PMP (Cycle Artemis Urban)

(a) Continuous A-PMP

ΔSOC	k_P		
	4	6	8
k_I 0	−0.244	−0.167	−0.125
0.1	0.007	0.007	0.007
0.2	-0.006	0.000	-0.001

m_f/m_f^*	k_P		
	4	6	8
k_I 0	1.00	1.00	1.00
0.1	1.01	1.01	1.01
0.2	1.01	1.01	1.00

(b) Discrete A-PMP

ΔSOC	k_P^d		
	2	4	8
T 15 s	0.044	−0.028	0.051
30 s	0.093	0.025	−0.002
60 s	0.034	0.093	0.016

m_f/m_f^*	k_P^d		
	2	4	8
T 15 s	1.02	1.02	1.04
30 s	1.04	1.03	1.01
60 s	1.05	1.03	1.02

The values shown are the total SOC variation and the fuel consumption, normalized with respect to the optimal value obtained with PMP

References

1. L. Serrao, A. Sciarretta, O. Grondin, A. Chasse, Y. Creff, D.D. Domenico, P. Pognant-Gros, C. Querel, L. Thibault, Open issues in supervisory control of hybrid electric vehicles: a unified approach using optimal control methods. Oil Gas Sci. Technol. Rev. IFP Energ. Nouv. (2013)
2. A. Sciarretta, L. Guzzella, Control of hybrid electric vehicles. IEEE Control Syst. Mag. **2**, 60–70 (2007)
3. L. Serrao, S. Onori, G. Rizzoni, A comparative analysis of energy management strategies for hybrid electric vehicles. ASME J. Dyn. Syst. Meas. Control **133**(3), 031012 (2011)
4. L. Serrao, S. Onori, G. Rizzoni, ECMS as a realization of Pontryagin's minimum principle for HEV control, in *Proceedings of the 2009 Conference on American Control Conference* (2009), pp. 3964–3969
5. C. Musardo, B. Staccia, Energy management strategies for hybrid electric vehicles Master's thesis (Politecnico di Milano, 2003)
6. C. Musardo, B. Staccia, S. Bittanti, Y. Guezennec, L. Guzzella, G. Rizzoni, An adaptive algorithm for hybrid electric vehicles energy management, in *Proceedings of the Fisita World Automotive Congress* (2004)
7. C. Musardo, B. Staccia, S. Midlam-Mohler, Y. Guezennec, G. Rizzoni, Supervisory control for NOx reduction of an HEV with a mixed-mode HCCI/CIDI engine, in *Proceedings of the 2005 American Control Conference* (2005)
8. C. Musardo, G. Rizzoni, Y. Guezennec, B. Staccia, A-ECMS: an adaptive algorithm for hybrid electric vehicle energy management. Eur. J. Control **11**(4–5), 509–524 (2005)
9. B. Gu, G. Rizzoni, An adaptive algorithm for hybrid electric vehicle energy management based on driving pattern recognition, in *Proceedings of the 2006 ASME International Mechanical Engineering Congress and Exposition* (2006)
10. A. Sciarretta, M. Back, L. Guzzella, Optimal control of parallel hybrid electric vehicles. IEEE Trans. Control Syst. Technol. **12**(3), 352–363 (2004)
11. D. Ambühl, L. Guzzella, Predictive reference signal generator for hybrid electric vehicles. IEEE Trans. Veh. Technol. **58**, 4730–4740 (2009)
12. L. Fu, Ü. Özgüner, P. Tulpule, V. Marano, Real-time energy management and sensitivity study for hybrid electric vehicles, in *Proceedings of 2011 American Control Conference* (2011)
13. S. Jeon, S. Jo, Y. Park, J. Lee, Multi-mode driving control of a parallel hybrid electric vehicle using driving pattern recognition. ASME J. Dyn. Syst. Meas. Control **124**, 141–149 (2002)
14. B. Gu, Supervisory control strategy development for a hybrid electric vehicle Master's thesis (The Ohio State University, 2006)
15. J. Kessels, M. Koot, P. van den Bosch, D. Kok, Online energy management for hybrid electric vehicles. IEEE Trans. Veh. Technol. **57**(6), 3428–3440 (2008)
16. A. Chasse, A. Sciarretta, J. Chauvin, Online optimal control of a parallel hybrid with costate adaptation, in *Proceedings of the 6th IFAC Symposium Advances in Automotive Control* (2010)
17. S. Onori, L. Serrao, G. Rizzoni, Adaptive equivalent consumption minimization strategy for hybrid electric vehicles, in *Proceedings of the 2010 ASME Dynamic Systems and Control Conference* (2010)
18. P. Tulpule, V. Marano, G. Rizzoni, Energy management for plug-in hybrid electric vehicles using equivalent consumption minimisation strategy. Int. J. Electr. Hybrid Veh. **2**(4), 329–350 (2010)
19. P. Khayyer, J. Wollaeger, S. Onori, V. Marano, U. Ozguner, G. Rizzoni, Analysis of impact factors for plug-in hybrid electric vehicles energy management, in *15th International IEEE Conference on Intelligent Transportation Systems (ITSC)* (2012), pp. 1061–1066

Chapter 8
Case Studies

8.1 Introduction

Two case studies are used to illustrate the concepts of modeling and control introduced in the book: a parallel hybrid vehicle (Sect. 8.2) and a power-split vehicle (Sect. 8.3). Details about the powertrain modeling, the definition of energy management problem, and the implementation of energy management strategies are provided for each case study. The vehicle itself is the same for both cases and its main characteristics are shown in Table 8.1.

8.2 Parallel Architecture

8.2.1 Powertrain Model

The first example of hybrid powertrain is the parallel architecture depicted in Fig. 8.1. The internal combustion engine and the electric motor are mechanically connected through a splitter box, in which their respective torque outputs are summed. A conventional 5-speed gearbox (GB) is present between the engine/motor and the wheels. A clutch is present to disconnect the engine from the gearbox input shaft during gear shift, and to allow electric-only driving. In the simulation model, the clutch is modeled as an ideal on/off component, neglecting the slipping phases, and the shifting operation is considered instantaneous. This is a full hybrid vehicle according to the categorization of Fig. 1.1, as it includes engine-only mode, electric-only mode, regenerative braking, and engine assist capabilities.

The kinematic chain composed of the gearbox and the differential/final drive (FD) introduces two gear ratios: the transmission ratio $g_{tr}(i_{tr})$, which is a function of the

© The Author(s) 2016
S. Onori et al., *Hybrid Electric Vehicles*, SpringerBriefs in Control,
Automation and Robotics, DOI 10.1007/978-1-4471-6781-5_8

Table 8.1 Vehicle parameters for the case studies [1, 2]

Frontal area, A_f	2.33 m²
Drag coefficient, C_d	0.26
Air density, ρ_{air}	1.22 kg/m³
Roll. resist. coeff, c_{roll}	0.024
Total vehicle mass, M	1370 kg
Wheel radius, R_{wh}	0.32 m
Final drive ratio, g_f	4.113
Distance of CG from front axle, a	1.2 m
Distance of CG from rear axle, b	1.5 m
Distance of CG from ground, h_{CG}	0.8 m

Fig. 8.1 Parallel HEV architecture

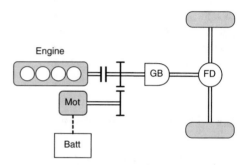

selected gear i_{tr}, and the constant final drive ratio g_{fd}. Thus, the torque T_{pwt} that reaches the wheels (see Sect. 2.4.4) is[1]

$$T_{pwt} = g_{tr}(i_{tr}) \cdot g_{fd} \cdot \left(T_{eng} + T_{mot} \right), \tag{8.1}$$

while the speed of the engine and motor is

$$\omega_{eng} = \omega_{mot} = \frac{v_{veh}}{R_{wh}} \cdot g_{tr}(i_{tr}) \cdot g_{fd}, \tag{8.2}$$

where v_{veh} is the vehicle speed and R_{wh} the wheel radius.

The torque curves and the efficiency maps of the engine and the electric motor are shown in Figs. 8.2 and 8.3, respectively.

---

[1]The dependence on t will not be explicitly indicated in the equations included in this chapter, for easier notation.

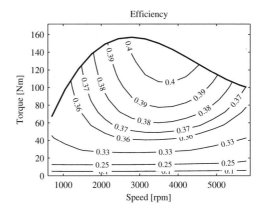

Fig. 8.2 Engine efficiency map (generic Diesel engine)

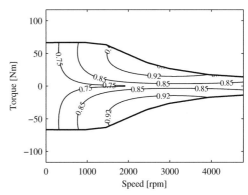

Fig. 8.3 Electric motor efficiency map (elaboration of data in [3])

Table 8.2 Battery parameters

Nominal charge capacity, Q_{nom}	6.5 Ah = 23400 C
Max voltage V_{max}	248 V
Energy capacity $E_{batt} = Q_{nom} V_{oc,nom}$	5.8 MJ = 1.6 kWh
Coulomb efficiency, η_{coul}	0.95
Max current (charge or discharge) I_{max}	130 A
Max power (charge or discharge) $P_{batt,max}$	31 kW

The battery is modeled using a zero-*th* order model, with SOC dynamics given by (2.33) (or (2.37)). A Li-ion battery pack is considered, with the data of Table 8.2 and the characteristics of open circuit voltage V_{oc} and internal resistance R_0 shown in Fig. 8.4.

Fig. 8.4 Battery characteristics, referred to the entire pack

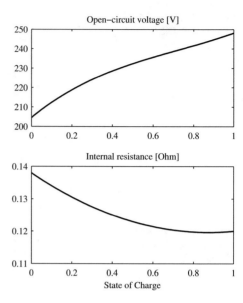

The objective is to minimize fuel consumption by sharing the total torque request between the internal combustion engine and the electric motor, maintaining charge-sustaining operation.

Cost function. The objective of minimization is the total fuel consumption, function of the engine torque T_{eng}, and speed ω_{eng}, expressed as total fuel energy:

$$J = Q_{lhv} \int_{t_0}^{t_f} \dot{m}_f(T_{eng}, \omega_{eng}) dt = \int_{t_0}^{t_f} P_{fuel}(T_{eng}, \omega_{eng}) dt. \tag{8.3}$$

The engine speed ω_{eng} is related to the vehicle speed v_{veh} and the gearbox ratio g_{tr} by (8.2). The gearbox ratio can be part of the optimization problem, or it can be taken as an external input. We consider the latter case, assuming that an independent transmission controller exists to select the gear based mainly on drivability considerations. The vehicle speed is considered as an external input for the energy management strategy, as is the total torque request T_{pwt} (or power request P_{req}), which is generated by the driver via the accelerator pedal or by the speed controller in simulation, according to the general control architecture shown in Fig. 3.2. From (8.1), if T_{pwt} and the gear index i_{tr} are imposed, the gearbox input torque is determined as

$$T_{gb} = T_{eng} + T_{mot} = \frac{T_{pwt}}{g_{tr}(i_{tr}) \cdot g_{fd}}. \tag{8.4}$$

As discussed in Sect. 3.4, the standard degree of freedom for the optimization problem is the battery power P_{batt}. It is directly and bi-univocally related to the motor torque, T_{mot}, because the motor speed is imposed by external inputs and the motor efficiency only depends on torque and speed (neglecting temperature effects). This allows using the motor torque as the control variable, which is more immediate for this powertrain architecture: in fact, given the torque of the electric motor, the engine torque is found by difference from T_{gb}:

$$T_{eng} = T_{gb} - T_{mot}. \tag{8.5}$$

The fuel power (proportional to the fuel consumption) is computed using the map in Fig. 8.2, so it is a function of engine speed and torque. Through (8.2) and (8.5), it is computed as a function of vehicle speed, total torque request, and motor torque as

$$P_{fuel} = Q_{lhv} \dot{m}_f(T_{eng}, \omega_{eng}) = P_{fuel}(T_{gb}, T_{mot}, v_{veh}). \tag{8.6}$$

System dynamics. The system dynamic equation represents the evolution of the battery state of charge as a function of the state itself and of the control input T_{mot} (or P_{batt}, for what said earlier). According to (2.37), the state of charge variation is

$$S\dot{O}C = -\frac{1}{\eta_{coul}^{sign(I(t))} Q_{nom}} \left[\frac{V_{oc}(SOC)}{2R_0(SOC)} - \sqrt{\left(\frac{V_{oc}(SOC)}{2R_0(SOC)}\right)^2 - \frac{P_{batt}}{R_0(SOC)}} \right] \tag{8.7}$$

with the battery parameters $V_{oc}(SOC)$ and $R_0(SOC)$ shown in Fig. 8.4.

It turns out it is more practical to use the electrochemical energy variation in place of state of charge as state in this case. According to the formulation introduced in Sect. 5.3.1:

$$E_{ech} = E_{batt} \cdot (SOC(t_0) - SOC(t)) \tag{8.8}$$

with $E_{batt} = V_{oc,nom} Q_{nom} \eta_{coul}$; the system dynamic equation is

$$\dot{E}_{ech} = P_{ech} = -E_{batt} S\dot{O}C. \tag{8.9}$$

The battery power is

$$P_{batt} = P_{em,e}(T_{mot}, \omega_{mot}), \tag{8.10}$$

where $P_{em,e}(T_{mot}, \omega_{mot})$ is the electrical power required by the electric machine to produce the torque T_{mot}, at the speed ω_{mot} (which is a function of the vehicle speed via the gearbox ratio i_{tr}). Therefore the state equation is a function only of SOC and P_{batt}, which in turn depends on the control input T_{mot}, and the external inputs v_{veh}, i_{tr}.

Control constraints. The value of engine and motor torque must remain within their respective limitations:

$$T_{mot,min}(\omega_{mot}) \leq T_{mot} \leq T_{mot,max}(\omega_{mot}), \tag{8.11}$$

$$T_{ice,min}(\omega_{eng}) \leq T_{eng} \leq T_{ice,max}(\omega_{eng}), \tag{8.12}$$

and, in addition, the motor power is saturated according to the minimum and maximum available electric power[2]:

$$P_{batt,min}(SOC) \leq P_{mot,e} \leq P_{batt,max}(SOC), \tag{8.13}$$

which is translated into an additional constraint on the control variable T_{mot}:

$$T'_{mot,min}(\omega_{mot}, P_{batt,min}) \leq T_{mot} \leq T'_{mot,max}(\omega_{mot}, P_{batt,max}) \tag{8.14}$$

PMP solution. Using the power-based formulation of Sect. 5.3.1, the Hamiltonian of the system has the form:

$$\begin{aligned} H(T_{mot}, & E_{ech}, T_{gb}, v_{veh}) \\ &= P_{fuel}\left(T_{mot}, T_{gb}, v_{veh}\right) + (\lambda + w(SOC)) \, P_{ech}\left(T_{mot}, E_{ech}, v_{veh}\right). \end{aligned} \tag{8.15}$$

The term $w(SOC)$ represents the additive penalty function (5.22) (see Fig. 5.3), active when state constraints are hit. In practice, it prevents the battery to be discharged when SOC is too low, by increasing the cost of positive P_{ech}, and it facilitates the battery to be discharged when SOC is too high by decreasing the cost of P_{ech}.

At each instant of time, the optimal solution $T_{mot}^*(t)$ is the one that minimizes the Hamiltonian function. One approach to minimization is to evaluate the function H for the complete set of admissible control values $T_{mot,vec} = \{T_{mot,min}, \ldots, T_{mot,max}\}$ at each time t, and then pick the one generating the lowest value of H. This optimal control value $T_{mot}^*(t)$ is applied to the system, which evolves generating the new values of state and external inputs. The co-state evolution is computed according to (5.30):

$$\dot{\lambda}(T_{mot}, E_{ech}, v_{veh}) = -\frac{\partial H}{\partial E_{ech}} = -\lambda \frac{\partial S\dot{O}C \, (P_{batt}, SOC)}{\partial SOC} \tag{8.16}$$

where the function $\frac{\partial S\dot{O}C}{\partial SOC}$ is computed numerically evaluating it from Eq. (8.7) and using the data in Fig. 8.4; its values as a function of P_{batt} and SOC are shown in Fig. 8.5.

---

[2]The maximum and minimum limitations on the power from/in the battery depend on battery state of charge (as well as temperature, although its effect is neglected in the models used here).

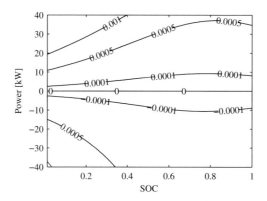

Fig. 8.5 The contour lines represent the values of the function $\frac{\partial S\dot{O}C(P_{batt},SOC)}{\partial SOC}$ in $[s^{-1}]$. The map and the contour lines are computed using the characteristics of Fig. 8.4

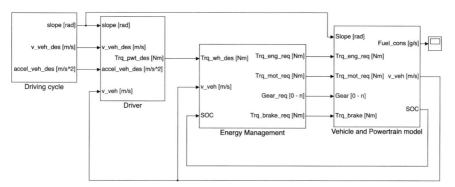

Fig. 8.6 Top-level view of vehicle simulator

8.2.3 Model Implementation

Figure 8.6 shows the implementation of the vehicle-level simulator which follows the structure defined in Fig. 2.4. The simulator is composed of the following main blocks:

- *Driving cycle* generates the sequence of setpoints for speed, acceleration, and slope that the vehicle should follow;
- *Driver* (Fig. 8.7) computes the torque setpoint necessary to follow the prescribed cycle; the setpoint is computed using (2.6), with an additional feedback term proportional to the speed tracking error (which may be different from zero when the cycle cannot be followed exactly due to powertrain actuator saturation);
- *Energy Management* generates the individual setpoints for the powertrain actuators (in this example relative to a parallel hybrid architecture, these are the engine and the electric motor). The inputs to the block are the total torque setpoint and the measurements from the vehicle, namely battery SOC and vehicle speed.

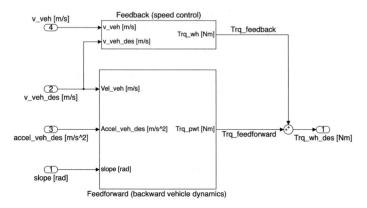

Fig. 8.7 *Driver* block of Fig. 8.6

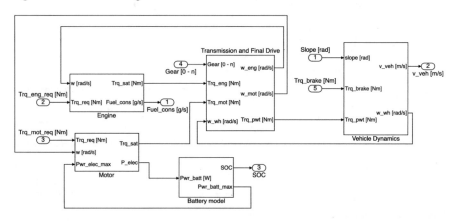

Fig. 8.8 *Vehicle and Powertrain model* block of Fig. 8.6

- *Vehicle and Powertrain model* (Fig. 8.8) contains the plant model, which takes the actuator setpoints as inputs and computes the evolution of vehicle speed and battery SOC, as well as the engine fuel consumption. The block includes the main powertrain components arranged according to a typical forward simulation approach, where vehicle speed is computed by integration of force according to (2.1) and Fig. 2.2. The speed is then fed back to all powertrain components. The engine and motor blocks saturate the torque demand according to torque and power limitations, and compute the consumption of fuel or electric power corresponding to the operating point, using static maps.

Figure 8.9 shows the implementation of the energy management using PMP; the function of each of the blocks composing the strategy can be detailed as follows:

- *Gear shifting strategy* implements a simple gear shifting strategy based on speed and torque thresholds: the highest gear compatible with the vehicle speed and with the driver's torque request is selected. This means that for a given vehicle speed,

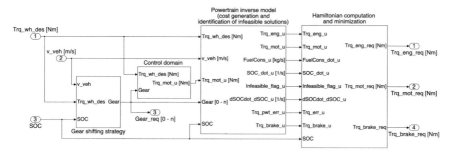

Fig. 8.9 *Energy Management Strategy* block of Fig. 8.6

the strategy selects the highest gear that allows the total torque output to match the driver's torque request, while keeping the engine speed in the allowable range (between idle and red-line).

The gear index, i_{tr}, is therefore computed using a rule-based algorithm that is outside of the optimal control strategy, for which the gear index is simply an external input. Note that the gear shifting strategy also takes SOC as an input, because the maximum torque of the motor depends on the battery state of charge as described by Eq. (8.13). Hence, the maximum output torque in a given gear is affected by SOC level.

- *Control domain* generates the set of control values for which the Hamiltonian is evaluated. This set is named here **Trq_mot_u** and is composed of N_u points, which include: $T_{mot} = 0$ (engine-only mode), $T_{mot} = T_{gb}$ (electric-only mode, or zero engine contribution), and then $N_u - 2$ values of torque distributed uniformly between the absolute minimum and maximum torque of the motor, to cover its entire torque range. In general, the suffix **_u** in the variable name denotes vectorial variables generated by this set of control candidates, i.e., arrays of size N_u. In the numerical examples that follow, $N_u = 22$.

- *Powertrain inverse model* implements the equations of the vehicle model described in Sect. 8.2.1, and outputs all the variables needed to compute the Hamiltonian function. These variables are arrays of size N_u, as they depend on the control input. Note that not all control candidates generate feasible solutions, since some of them may not meet all the instantaneous constraints (e.g., some motor torque values may exceed battery power limitations, others may correspond to infeasible engine torque values, etc.). In order to exclude the infeasible solutions, the variable **Infeasible_flag_u** is created, which contains a flag identifying infeasible solutions, i.e., solutions that do not meet the control constraints.[3] These infeasible solutions have a very large cost associated to them, in order to be excluded from the ensuing minimization.

- *Hamiltonian computation and minimization* computes the Hamiltonian function for all elements in the control arrays, and then identifies the index of the array

---

[3] **Infeasible_flag_u** is an array of size N_u composed of zeros and ones: zeros for the solutions that meet all the constraints, ones for those that do not meet some of them.

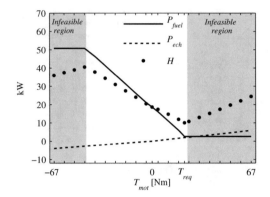

Fig. 8.10 Example of Hamiltonian computation and minimization. The discretized control variable is the motor torque T_{mot}; the *solid line* represents the values of engine fuel power P_{fuel} and the *dashed line* shows the values of the battery electrochemical power P_{ech}. The resulting Hamiltonian H computed for each control candidate value is indicated by the *dots*. The *gray* areas represent values of the control corresponding to infeasible solutions. In this example, the minimum value of Hamiltonian corresponds to $T_{mot} = T_{req}$, i.e., to electric-only propulsion

corresponding to the minimum value. The index is used to select, from the arrays of engine, motor, and brake torque, the optimal values that are then actuated in the plant. A numerical example of the values taken at a given time by the Hamiltonian function and its constituents is shown in Fig. 8.10, where the full control range is represented.

8.2.4   Simulation Results

The optimal solution is obtained by applying PMP and using the shooting method to find the optimal initial value of the co-state variable. The results of the iterative search procedure, performed for a Worldwide harmonized Light vehicle Test Procedure (WLTP) driving cycle [4], are shown in Figs. 8.11 and 8.12: starting from an initial guess for λ_0, the problem is solved and the final state of charge value, $SOC(t_f)$ is compared to the target, SOC_{target}. Depending on the difference $SOC(t_f) - SOC_{target}$ the value of λ_0 is increased or decreased in the next iteration, and the driving cycle is simulated again with a new initial value λ_0. At the generic nth iteration, the value of λ_0 is set to

$$\lambda_0(n) = \frac{1}{2} \left(\lambda_{inf}(n-1) + \lambda_{sup}(n-1) \right) \tag{8.17}$$

where λ_{inf} and λ_{sup} are two variables introduced to implement a bisection method [5]. After being initialized at arbitrary values, λ_{inf} and λ_{sup} are updated at each step

Fig. 8.11 Shooting method: convergence toward the optimal initial co-state value

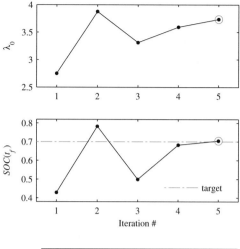

Fig. 8.12 Shooting method: iterations shown in the plane $(\lambda_0, (SOC(t_f)))$

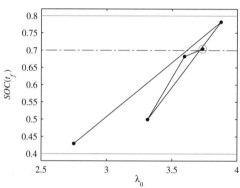

of the iteration according to the following rules:

$$\text{if } SOC(t_f) - SOC_{target} < 0 : \begin{cases} \lambda_{inf}(n) = \lambda_0(n-1) \\ \lambda_{sup}(n) = \lambda_{sup}(n-1) \end{cases} \qquad (8.18)$$

$$\text{if } SOC(t_f) - SOC_{target} > 0 : \begin{cases} \lambda_{inf}(n) = \lambda_{inf}(n-1) \\ \lambda_{sup}(n) = \lambda_0(n-1) \end{cases} \qquad (8.19)$$

In the example shown, the initial values at $n = 0$ are $\lambda_{inf} = 0.5$ and $\lambda_{sup} = 5$; the search terminates when $|SOC(t_f) - SOC_{target}| < 0.01$. The convergence of the bi-section method is reached in 5 iterations, as shown in Figs. 8.11 and 8.12.

The bisection method is applied to two PMP formulations: (1) dynamic co-state expressed according to (8.16), and (2) constant co-state where λ is kept at the value

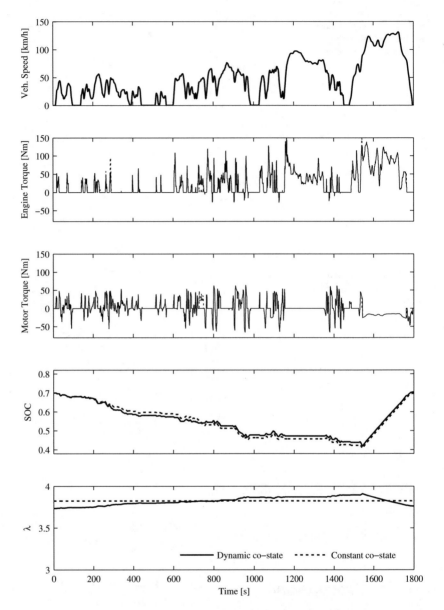

Fig. 8.13 Optimal solutions for cycle WLTP, obtained solving PMP with: (1) dynamic co-state and (2) constant co-state. The reduction of fuel consumption with respect to the corresponding conventional vehicle (identical, but without electric motor) is 21.5 % for both cases. Note how the optimal control policy makes the motor torque negative in the last part of the cycle, thus using the engine to recharge the battery up to the target final *SOC*

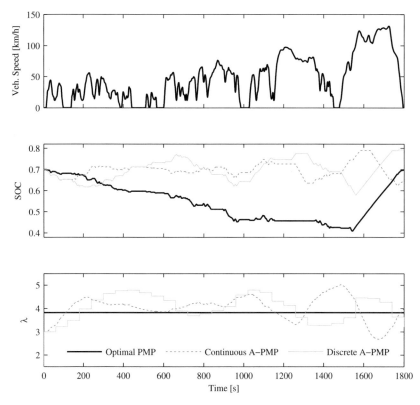

Fig. 8.14 Optimal solution (with constant co-state) compared to adaptive strategies (7.1) and (7.2). The continuous A-PMP parameters are $k_P = 4$, $k_I = 0.2$; the discrete A-PMP parameters are $k_P^d = 8$, $T = 60$ s. After SOC correction, the fuel consumption increase with respect to the optimal solution is 2.4 % with continuous A-PMP, and 2.1 % with discrete A-PMP

λ_0 for the entire driving cycle. Figure 8.13 shows a comparison of the two cases (each with the optimal λ_0 computed from the iterative search).

Figure 8.14, instead, compares the optimal solution to the results of the two adaptive strategies introduced in Sect. 7.3.1, i.e., the continuous and the discrete A-PMP.

8.3 Power-Split Architecture

8.3.1 Powertrain Model

The example of powertrain modeling presented in this section is based on the Toyota Hybrid Synergy Drive (HSD) [3, 6]. Being the first successful hybrid electric technology on the market, this system has been extensively studied in the literature

Fig. 8.15 Power split
architecture with planetary
gear train arrangement

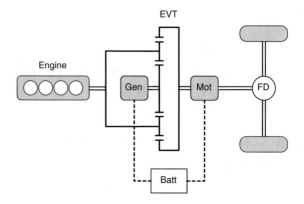

[1, 2, 7–9]. The hybrid architecture consists of an electrically variable transmission (EVT), composed of a planetary gear set to which the engine and one electric machine (the generator) are connected, as shown in Fig. 8.15; a second, more powerful electric motor is connected to the EVT output (the ring gear). The battery pack provides electrical power to the two electric machines.

The engine is connected to the carrier shaft of the planetary gear set[4] as shown in Fig. 8.15; the generator is connected to the sun, while the ring is connected to the output shaft. The motor is also connected to the output shaft, thus the motor and the ring drive the powertrain output together. A quasi-static modeling approach is used for energy analysis, neglecting the inertia and the dynamics of engine, electric machines, and all gears and shafts.

The torque at the wheels is

$$T_{pwt} = g_{fd} \cdot (T_r + T_{mot}) \tag{8.20}$$

where T_{mot} is the motor torque, T_r is the planetary ring torque, and g_{fd} is the final drive ratio. T_r is related to the generator and motor torque by the general planetary gear set equation (2.23):

$$T_{eng} = T_c = (1 + \rho)T_r \tag{8.21}$$

$$T_{gen} = T_s = \rho \cdot T_r \tag{8.22}$$

where $\rho = N_s/N_r$ is the planetary gear ratio ($N_r = 78$ and $N_s = 30$ [2]).

Given (8.21) and (8.22), one of T_r, T_{eng}, and T_{gen} is sufficient to determine the other two.

---

[4]see also Fig. 2.11.

The kinematic constraint (2.21) can be written in this case as

$$(1 + \rho)\omega_{eng} = \rho\omega_{gen} + \omega_{mot}. \tag{8.23}$$

The motor speed (which is equal to the ring speed) is proportional to the wheel speed, since there is a fixed gear (the differential) between the ring/motor shaft and the wheels; therefore, it is also proportional to the vehicle longitudinal speed:

$$\omega_{mot} = \omega_r = g_{fd}\frac{v_{veh}}{R_{wh}} \tag{8.24}$$

where v_{veh} is the vehicle speed and R_{wh} the wheel radius. Using (8.23), the engine speed ω_{eng} can be related to the generator speed ω_{gen} and the vehicle speed as follows:

$$\omega_{eng} = \frac{\rho}{1+\rho}\omega_{gen} + \frac{g_{fd}}{1+\rho}\frac{v_{veh}}{R_{wh}}. \tag{8.25}$$

Equation (8.25) shows an interesting characteristic of this powertrain: the engine speed can be made to assume any value (within the admissible range) independently from the vehicle speed, by varying the speed of the generator. This is the reason why this kind of arrangement is also defined as electrically continuously variable transmission (E-CVT), pointing out the similarity with the CVT technology used in conventional vehicles. In fact, both realize a transmission with no fixed ratios, but rather a continuously varying ratio between the engine speed and the vehicle speed, as shown by the graphical representation of (8.25) in Fig. 8.16.

By varying the generator speed, it is therefore possible to keep the engine in the maximum efficiency range for each torque level.

The battery is the same as that described in Table 8.2; the engine, motor, and generator maps are shown in Figs. 8.17, 8.18, and 8.19 respectively.

Fig. 8.16 EVT ratio: Engine speed versus vehicle speed, for several values of generator speed (admissible engine range in *bold*)

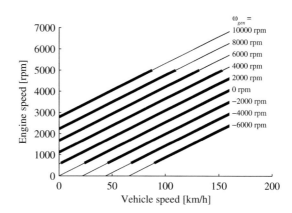

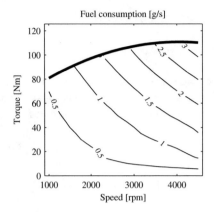
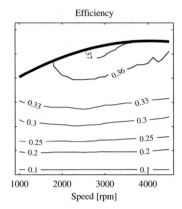

Fig. 8.17 Engine maps [1]

Fig. 8.18 Motor map
(elaboration of data from [3])

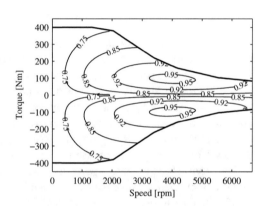

Fig. 8.19 Generator map
(elaboration of data from [3])

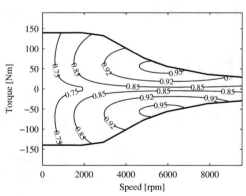

8.3.2 Optimal Control Problem Solution

The energy management problem formulation and the definition of the constraints follow the same pattern as in the previous case study (Sect. 8.2.2). The differences between the parallel and powersplit architecture are in kinematic equations and the different control variables, as described in the following.

The required torque at the wheels T_{pwt} is obtained from (8.20), while the ring torque T_r is computed from (8.21) as a function of the engine torque T_{eng}. The generator torque is determined by (8.22). Thus, the motor and engine torques determine the amount of torque transmitted at the wheels, while the generator speed is set in order to let the engine operate at high efficiency. The energy management problem has two degrees of freedom (once the output torque T_{pwt} is imposed): the engine torque T_{eng} and the generator speed ω_{gen}. Together, they define the battery power, casting once again the problem in the general framework defined in Sect. 3.4.

The generator mechanical power can be related to the engine torque and speed using (8.21) and (8.22):

$$P_{gen,m} = \omega_{gen} T_{gen} = \omega_{gen} \frac{\rho}{1 + \rho} T_{eng}, \qquad (8.26)$$

while the motor power is

$$P_{mot,m} = \omega_{mot} T_{mot}. \qquad (8.27)$$

The total electric power at the battery is

$$P_{batt} = P_{mot,e} + P_{gen,e} \qquad (8.28)$$

where the electric power of motor and generator is related to their mechanical power by the efficiency of each machine:

$$P_{gen,e} = \begin{cases} \eta_{gen} P_{gen,m} & \text{if } P_{gen,m} < 0 \\ \frac{1}{\eta_{gen}} P_{gen,m} & \text{if } P_{gen,m} \geq 0 \end{cases} \qquad (8.29)$$

$$P_{mot,e} = \begin{cases} \eta_{mot} P_{mot,m} & \text{if } P_{mot,m} < 0 \\ \frac{1}{\eta_{mot}} P_{mot,m} & \text{if } P_{mot,m} \geq 0. \end{cases} \qquad (8.30)$$

8.3.3 Model Implementation

The simulator implementation is the same as the previous case study, but in this case the *Energy Management Strategy* has the structure of Fig. 8.20: the control array is composed of permutations of the two control variables, T_{eng} and ω_{gen}, thus

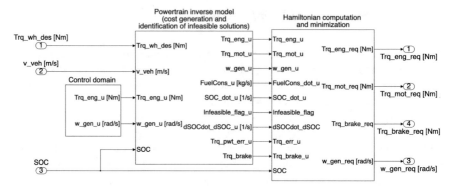

Fig. 8.20 *Energy Management Strategy* block for EVT case study

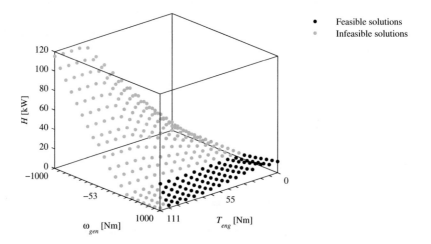

Fig. 8.21 Hamiltonian function computed over the discretised control space, T_{eng} and ω_{gen}, at a given time instant

increasing the overall number of candidates to be evaluated (for instance, if each of the two variables can take one of 20 values, the total number of candidates is 400). The Hamiltonian function at each instant is therefore a 2-D surface, computed for all combinations of the two control variables shown in Fig. 8.21.

8.3.4　Simulation Results

The optimal solution obtained with PMP after optimization of the co-state is shown in Fig. 8.22, for an urban driving cycle. The corresponding engine operating points resulting from the optimization are reported in Fig. 8.23, showing how the control

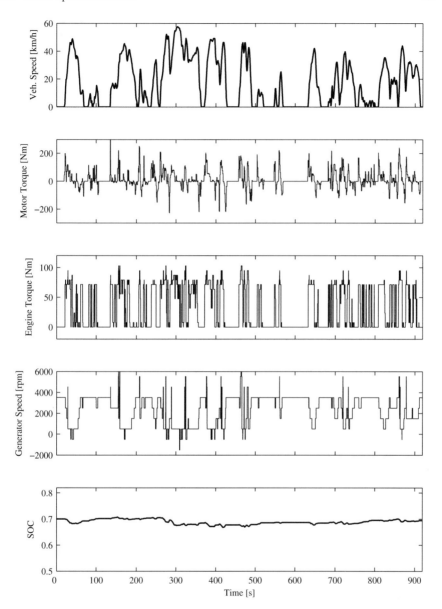

Fig. 8.22 Optimal solutions obtained solving PMP with constant co-state (Cycle Artemis Urban). The optimal co-state value is found to be $\lambda = 2.504$

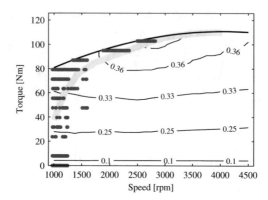

Fig. 8.23 Engine operating points corresponding to the solution of Fig. 8.22; the *gray line* indicates the optimal operation line (maximum engine efficiency)

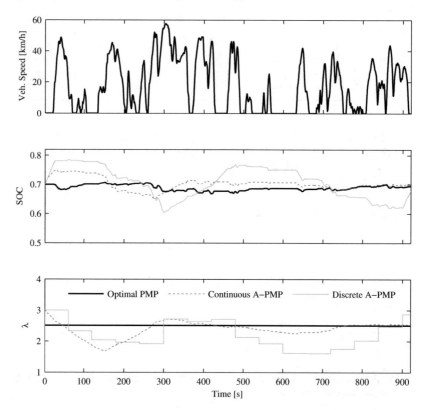

Fig. 8.24 Optimal solution of Fig. 8.22 compared to adaptive strategies (7.1) and (7.2). The continuous A-PMP parameters are $k_P = 4, k_I = 0.2$; the discrete A-PMP parameters are $k_P^d = 8, T = 60\,\mathrm{s}$. After SOC correction, the fuel consumption increase with respect to the optimal solution is 1.6 % with continuous A-PMP, and 7.8 % with discrete A-PMP

leads to the selection of engine points around the optimal operation line. Finally, the same optimal solution is compared to the adaptive strategies in Fig. 8.24. Unlike the example proposed in Fig. 8.14, in this case the discrete A-PMP is sensibly worse than the optimal solution and the continuous A-PMP: the reason is that the *SOC* reaches the upper bound in several instances, which prevents the recuperation of all available regenerative braking energy, therefore reducing the overall performance. In addition, the optimal solution makes use of the EVT ability to recirculate power between motor and generator, but uses a limited amount of battery power, as seen by the small variation of *SOC* in Fig. 8.22; on the other hand, the adaptive strategies, without a priori knowledge of the optimal co-state value, show greater *SOC* variations due to more usage of the battery, resulting in a small penalty in fuel consumption due to the extra charge–discharge losses.

References

1. J. Meisel, An analytic foundation for the Toyota Prius THS-II powertrain with a comparison to a strong parallel hybrid-electric powertrain, SAE paper 2006-01-0666 (2006)
2. C.W. Ayers, J.S. Hsu, L.D. Marlino, C.W. Miller, G.W. Ott, C.B. Oland, Evaluation of 2004 Toyota Prius hybrid electric drive system. Technical report, Oak Ridge National Laboratory (2004)
3. T.A. Burress, S.L. Campbell, C.L. Coomer, C.W. Ayers, A.A. Wereszczak, J.P. Cunningham, L.D. Marlino, L.E. Seiber, H.T. Lin, Evaluation of the 2010 Toyota Prius hybrid synergy drive system. Technical report, Oak Ridge National Laboratory (2011)
4. Worldwide harmonized light vehicles test procedures. https://www2.unece.org/wiki/pages/viewpage.action?pageId=2523179
5. R.L. Burden, F.J. Douglas, *Numerical Analysis* (PWS Publishers, 1985)
6. Toyota hybrid system THS II, Toyota motor corporation. Technical report (2003)
7. T. Hofman, R. van Druten, A. Serrarens, J. van Baalen, A fundamental case study on the Prius and IMA drivetrain concepts. Technical report, Technische Universiteit Eindhoven, The Netherlands (2015)
8. J. Meisel, An analytic foundation for the two-mode hybrid-electric powertrain with a comparison to the single-mode Toyota Prius THS-II powertrain. SAE Paper 2009-01-1321 (2009)
9. J. Liu, H. Peng, Z. Filipi, Modeling and analysis of the Toyota Hybrid System, in *Proceedings of the 2005 IEEE/ASME International Conference on Advanced Intelligent Mechatronics* (2005), pp. 134–139

Series Editors' Biographies

Tamer Başar is with the University of Illinois at Urbana-Champaign, where he holds the academic positions of Swanlund Endowed Chair, Center for Advanced Study Professor of Electrical and Computer Engineering, Research Professor at the Coordinated Science Laboratory, and Research Professor at the Information Trust Institute. He received the B.S.E.E. degree from Robert College, Istanbul, and the M.S., M.Phil., and Ph.D. degrees from Yale University. He has published extensively in systems, control, communications, and dynamic games, and has current research interests that address fundamental issues in these areas along with applications such as formation in adversarial environments, network security, resilience in cyber-physical systems, and pricing in networks.

In addition to his editorial involvement with these Briefs, Basar is also the Editor-in-Chief of Automatica, Editor of two Birkhäuser Series on Systems & Control and Static & Dynamic Game Theory, the Managing Editor of the Annals of the International Society of Dynamic Games (ISDG), and member of editorial and advisory boards of several international journals in control, wireless networks, and applied mathematics. He has received several awards and recognitions over the years, among which are the Medal of Science of Turkey (1993); Bode Lecture Prize (2004) of IEEE CSS; Quazza Medal (2005) of IFAC; Bellman Control Heritage Award (2006) of AACC; and Isaacs Award (2010) of ISDG. He is a member of the US National Academy of Engineering, Fellow of IEEE and IFAC, Council Member of IFAC (2011–2014), a past president of CSS, the founding president of ISDG, and president of AACC (2010–2011).

Antonio Bicchi is Professor of Automatic Control and Robotics at the University of Pisa. He graduated from the University of Bologna in 1988 and was a postdoc scholar at M.I.T. A.I. Lab between 1988 and 1990. His main research interests are in:

- dynamics, kinematics and control of complex mechanical systems, including robots, autonomous vehicles, and automotive systems;
- haptics and dextrous manipulation; and
- theory and control of nonlinear systems, in particular hybrid (logic/dynamic, symbol/signal) systems.

© The Author(s) 2016   111
S. Onori et al., *Hybrid Electric Vehicles*, SpringerBriefs in Control,
Automation and Robotics, DOI 10.1007/978-1-4471-6781-5

He has published more than 300 papers in international journals, books, and refereed conferences.

Professor Bicchi currently serves as the Director of the Interdepartmental Research Center "E. Piaggio" of the University of Pisa, and President of the Italian Association or Researchers in Automatic Control. He has served as Editor-in-Chief of the Conference Editorial Board for the IEEE Robotics and Automation Society (RAS), and as Vice President of IEEE RAS, Distinguished Lecturer, and Editor for several scientific journals including the *International Journal of Robotics Research, the IEEE Transactions on Robotics and Automation, and IEEE RAS Magazine.* He has organized and co-chaired the first WorldHaptics Conference (2005), and Hybrid Systems: Computation and Control (2007). He is the recipient of several best paper awards at various conferences, and of an Advanced Grant from the European Research Council. Antonio Bicchi has been an IEEE Fellow since 2005.

Miroslav Krstic holds the Daniel L. Alspach chair and is the founding director of the Cymer Center for Control Systems and Dynamics at University of California, San Diego. He is a recipient of the PECASE, NSF Career, and ONR Young Investigator Awards, as well as the Axelby and Schuck Paper Prizes. Professor Krstic was the first recipient of the UCSD Research Award in the area of engineering and has held the Russell Severance Springer Distinguished Visiting Professorship at UC Berkeley and the Harold W. Sorenson Distinguished Professorship at UCSD. He is a Fellow of IEEE and IFAC. Professor Krstic serves as Senior Editor for *Automatica and IEEE Transactions on Automatic Control* and as Editor for the Springer series *Communications and Control Engineering.* He has served as Vice President for Technical Activities of the IEEE Control Systems Society. Krstic has co-authored eight books on adaptive, nonlinear, and stochastic control, extremum seeking, control of PDE systems including turbulent flows and control of delay systems.

Printed in the United States
By Bookmasters